LIFE ON MARS

LIFE
ON
MARS

COMPILED BY
Robert A. Granger
Professor
Department of Mechanical Engineering
U.S. Naval Academy
Annapolis, MD

SCIENTIFIC ARCHIVES, INC.
Lady Lake, Florida • Zurich

To my daughter, Erin

Published by the Scientific Archives, Inc., corporate headquarters
1308 Santa Maria Avenue, Lady Lake, FL 32159. Publication and
Merchandising Division at P.O. Box 600, Riva, MD 21140.

First published 1997

Published in the United States of America

Library of Congress Catalog Card Number: 97-069448

Granger, Robert A.
    Life on Mars/Compiled by Robert A. Granger.
    176 p.   cm.—(Scientific Archives Series in Science)
    "A Scientific Archives publication"
    Bibliography: p.
    Includes index
    ISBN 0-9659643-0-2
    1. Mars—geophysics. I. Title.
    Book file catalog #

Printed in the United States of America

10 9 8 7 6 5 4 3 2 1

CONTENTS

PREFACE

Public excitement fueled by possible evidence of life on Mars recently found in a fossilized meteorite ALH84001 is greater today than ever before. However, the outlook is dim that earthlings may soon garner some truth about "life on Mars." This is largely due to the enormous expense involved in searching for answers.

In November 1996, a seven-ton spacecraft filled with 22 instruments necessary to conduct a search for life on Mars crashed into the Pacific. It will take many years for the European and Soviet space agencies to recover from the loss, pushing back an important aspect of the quest for knowledge. NASA has recently suffered significant budget cuts that will surely curtail the success of future missions, including a scheduled launch in 2005 to gather and return Martian rocks to earth for analysis.

We are reaching a crisis in the space program when reasons for exploring our solar system are threatened. Some threats come from the scientific community, some are political, but most are economical. In the past there was widespread support that we needed to grasp the knowledge of the origin of not only our planetary system but others, and that we needed to fully understand the limitations of an environment that can sustain life. It is folly, it is naive, to believe or to be content with the belief that life is guaranteed and assured for the next one, ten, or a hundred million years to come. There are too many asteroids out in space that threaten earth; even worse, they are in our solar system. Of course they come in assorted sizes, but any one of the larger asteroids could destroy hundreds of millions of human beings as well as eliminate untold specie of flora and fauna if it entered our atmosphere. All we need do is recall what happened to the dinosaur. We need, therefore, to search for an alternate planet in which we might continue to survive in the event the earth is doomed.

In the search for life outside of our own planet, we begin with the planet closest to us, one that is not too dissimilar from our own. That is Mars. Some scientists argue for an immediate course of action in our search for an alternate planet; a crash program to send humans to Mars. Others urge caution, as the quest must be logical, continuous and thorough, so that minimum errors occur, any one of which has the potential to invalidate any interim conclusions. Thus, machines, free of earth contaminates, could function to a remarkable degree in a Martian environment. They could perform tests, then telemeter the data to earth for analysis. The Clinton administration is sympathetic to this line of reason. Vice President Gore, who oversees NASA's policies, is urging the U.S. pursue a "robust" space program. There are also many friends in the Congress who know we must not stifle our space research.

But space exploration is expensive: over $200 million for a small robotic planetary mission. With only $5.5 billion allocated this year to run all the projects at NASA, there is little optimism to see how NASA can divert funds from the enormously popular space station to planetary science. All the space programs are behind schedule, all are under funded, and so if money has to be spent to beef up the space station, it will probably come from the space science programs.

The purpose of this book is to inform the reader of the background and ideas that have gone into the search for life on Mars. Much of the technical explanations have been confined in select technical journals that seldom reach beyond those scientists involved in the Martian studies. It is my intent to present the technical aspects in a fashion the reader can understand so that the full picture of the search for life on Mars is comprehensible. Thus, even high school students reading this work would be able to follow the narrative without understanding any of the mathematical models.

This book could not have been written without the results from so many scientists in nearly every aspect of geophysics, planetary physics, biophysics, chemistry, and engineering. To cite all of the contributors would be exhaustive. The ideas contained herein are based solely on scientific inquiry, and the interpretations have been credited wherever possible. Thus, I have only compiled some of the vast information, not authored or originated the findings. Conclusions that are stated are based on scientific interpretations, and thus they are prone to dispute. Scientists try to avoid making conclusions factual, always leaving enough in the description to amend facts as being simply strong possibilities.

The research embraced in this book took place over a span of eight years, commencing at Yale University while on leave of absence from the U.S. Naval Academy. Thousands of papers were digested to gain insight into this fascinating subject. Acknowledgment is made to NASA for the Mars photographs, Professors Steve Squyres and Carl Sagan of Cornell for reviewing much of this work, scientists of the U.S. Geological Survey for fascinating discussions, Doris Keating of TechnoGraph Productions for the editing and typing services, the staff at Scientific Archives, Inc., and especially my wife, Ruth, who sacrificed the "fun things we do together" while working evenings and weekends on this project.

ROBERT GRANGER

Annapolis, Maryland
June 1997

1

INTRODUCTION

1.1 BACKGROUND

This book is a treatise dealing with one of the classical mysteries today: namely, is there or has there ever been any life on the planet Mars? The only positive answer to this question will come with landing man on Mars. Since this may not take place for another decade, we can make certain qualified estimates using carefully controlled scientific measurements. Scientists have made spectacular evaluations in the past 15 years based on studying photographs and technical data from two important space flight missions. These scientists presented their findings at international symposiums as well as published their technical papers in refereed international scientific journals. By the way, this was after considerable scrutiny by experts in their field. Every pitfall, each assumption and premise was meticulously examined. Experimental errors, bias, and appropriateness of reference datum were questioned. Many of the experimental conclusions were supported by elegant mathematical models lending further credence to their conclusions. These models also had imposed assumptions necessary to make the theory tractable and solvable.

Water on Mars plays perhaps the single most important role in creating and sustaining life. Obviously without it, man has a limited time on Mars. But if there is water, be it surface ice, subterranean liquid water, or water vapor, then man can pursue the possibility of establishing colonies on Mars.

Mars is the third smallest planet, the fourth planet from our Sun, and has evoked the curiosity of scientists and writers for centuries largely because of the possibility that life had or might exist there. After the invention of the telescope in the early seventeenth century, astronomers discovered surface markings on Mars's surface that were attributed to canals and seas. Initially they were thought to be optical illusions. Canals conjured up images of intelligent life, and novelists took it from there. Additional studies by astronomers showed Mars's rotation was similar to Earth's. It had polar caps, clouds, and storms, soaring volcanoes[*][(1)] and huge rift valleys[(2)]. It was revealed that Mars is more earth-like than any of the other planets in our solar system, so it is a natural quest to wish to explore it, despite its hostile environment.

For 300 years, Mars was viewed as earth-like and as able to support life. Up until fairly recently, many conjectured the life forms as humanistic, though some thought it would be miraculous if the life forms were larger than microorganisms. Telescopes were not powerful enough to obtain the high resolution to detect life, so the only recourse was to go to Mars. With the advent of rocketry (which evolved as a machine for war) there was now the means.

In the eyes of much of the general public, the space age began on 4 October 1957, when the first artificial Earth satellite, Sputnik I, was launched into orbit by the Soviet Union. In reality, however, visions of the conquest of space have been with us since 4,000 years before the birth of Jesus. The first firework rockets were used by the Chinese about that time, and Babylonian paintings from that era show human figures in skyward flight. Rockets were used by the British against the United States in the siege of Baltimore in 1814, prompting Francis Scott Key to write "by the rockets' red glare" in The Start Spangled Banner.

Major contributions to the development of space flight were made in the early part of the twentieth century by Karstantin Tsiolkovsky in Russia, Herman Oberth in Germany, and Robert H. Goddard in the United States. In March 1926, Goddard successfully flew the first liquid fueled rocket.

During World War II, rocket development in the U.S. was not a major effort. America concentrated on jet-assisted airplane take-offs and sounding rockets. In Germany, the Rocket Development Center at Peenemunde was a major effort. It developed first the V-1 "buzz-bomb" and then the V-2, the first operational military rocket and the forerunner of all the satellite launch vehicles and ballistic missiles that exist today.

---

[*] All footnotes are given in Notes, page 150, of this book.

Following the end of World War II, the Naval Research Laboratory (NRL) was one of several U.S. laboratories that obtained captured V-2 rockets. NRL began a program of upper atmosphere research and in October 1946 obtained the first solar spectrum that went to wavelengths below the atmospheric ozone absorption limit. Between 1946 and 1952 NRL launched 63 V-2 rockets with a success rate of over 60% in a series of measurements of upper atmosphere temperature, pressure, and winds.

In 1955 the White House announced that the U.S. would launch a satellite as part of the country's participation in the International Geophysical Year (IGY), July 1957 to December 1958. NRL was given the responsibility for the U.S. IGY satellite project, which was called Project Vanguard.

In comparison to the Soviet Sputnik program, Project Vanguard was a failure. Out of 11 launch attempts between 6 December 1957 and 18 September 1959, only three satellites reached orbit successfully. The television coverage of the rocket crumbling to earth in the first launch attempt of December 1957, just two months after Sputnik, galvanized national support for the infant space program.

On 31 January 1958, the Army Ballistic Missile Agency—Jet Propulsion Laboratory's Explorer threw America's first satellite into an earth's orbit. It weighed 14 kg compared to 500 kg of Sputnik 2 which had gone up earlier. America had lost the race to be first in space. In March 1958, America needed a strong national space program. President Dwight Eisenhower, Senator Lyndon Johnson and others in Congress were united in wanting to place the space program in a peaceful, research-oriented atmosphere. With the military aspects under Department of Defense (DOD) control, President Eisenhower signed into law P.L. 85-568, the National Aeronautics and Space Act of 1958. This law set into being civilian aeronautical and space research. The National Aeronautics and Space Administration (NASA) came into existence out of this act on 1 October 1958. A site in Beltsville, Maryland, was selected for a new center for space science research, satellite development, flight operation, and tracking. In March 1961 it was named the Robert H. Goddard Space Flight Center.

1.2 SPACE POLICY DOCTRINE AND LAWS

Any vehicle (ship, aircraft, rocket) moving through any fluid (liquid, gas) in international or free space must be under the influence of international laws to prevent exploitation and abuse. In this context, there must be both national and international laws.

1.2.1 U.S. Space Policy and Laws

The launch of Sputnik I by the Soviet Union alerted the United States that it was number two in the space race. That event which was mentioned earlier produced the first U.S. space policy: the National Aeronautics and Space Act of 1958. Although the Act has had numerous amendments, its guidelines remain unchanged, and it forms the basis of the U.S. space policy. Since national interests are in a state of flux, the U.S. space policy undergoes continuous revision. It is important to discuss specific aspects of our space policy since it governs the Mars space program.

The emphasis of the national policy is that (1) the U.S. space program is to be of benefit to the <u>security</u> and general welfare of the United States <u>and</u> to all mankind. The activities associated with the development of weapons systems, military operations, or the defense of the United States were to be the responsibility of DOD with the President determining jurisdiction in borderline cases. The space program was to be the responsibility of a civilian agency (NASA) except in the area of military security; however, effective coordination among all involved (civilian and military organizations) was to be maintained at all times. (2) Overall policy direction was to come from a high level council chaired by the President. (3) Congressional oversight was to be carried on by two newly created standing committees. (4) Any urgency of action and long-range objectives sought would be determined (unofficially) by the Soviet challenge.

Under the Act, NASA was charged with the responsibility to investigate outer space and to report its results. For example, in the commercialization of space, each executive administration had its own policy:

- President Eisenhower: "To achieve the early establishment of a communication satellite system which can be used on a commercial basis is a national objective which will require both the concerted capabilities and funds of both Government and private enterprises."

- President Kennedy: "...accelerating the use of satellites for worldwide communications ...[and]...a satellite system for worldwide weather observation."

- President Nixon: Nixon indirectly supported commercialization of space by starting the Space Shuttle Program.

- President Carter: Carter had nine points on the policy of commercialization of space: i) to pursue scientific knowledge and develop useful commercial and governmental applications of space, ii) to hold that space systems of any nation are national property and have the right of passage without interference, iii) to reject any claims of sovereignty over outer space, iv) to encourage domestic commercial exploitation of space resources for economic benefit, v) to develop a fully operational Space

Transportation System (STS) through NASA in cooperation with DOD, vi) to provide the private sector to take greater responsibility, vii) to develop a legal regime for space to assure safe and peaceful use, viii) to involve the LANDSAT system with the private sector, and ix) to prepare a plan to encourage private investment in the exploitation of space.

- President Reagan: i) to expand the private sector to invest and become involved in space, ii) to assure the space shuttle's utility to civilian users, and iii) to develop STS to meet national needs and to be available to authorized domestic and foreign users.

- President Bush: Bush called for expanding human presence and activity beyond earth's orbit into the Solar System; obtaining scientific, technological and economic benefits for the American people; encouraging private sector participation in space; improving the quality of life on earth; strengthening national security; and promoting international cooperation in space. In August 1989 NASA began an extensive review to summarize the technology and strategies for going back to the Moon and on to Mars.

- President Clinton: Clinton's space policy had nothing dramatically different from President Bush's space policy. The Clinton space policy is a shift away from the Cold War mentality of a "Space Race" with the former Soviet Union. The 1996 policy does not require U.S. preeminence in any space area. The policy allows other nations to take the lead and initiative in its stead. There is no wording that supports an expanding human presence in space. The policy clearly states that the U.S. will conduct satellite photo reconnaissance; that the U.S. government agencies will purchase commercially available goods and services; that it will seek mechanisms to stimulate private-sector investments and ownership of space assets. Details of the U.S. space policy are obtained on the web site: http:\\www.whitehouse.gov\WH\EOP\OSTP\NSTC\html/fs/fs-5.html.

Thus, each administration allowed the space program to grow in accordance with its space policy.

1.2.2 International Laws

In the late 1950s, President Eisenhower encouraged the United Nations to take action to preserve outer space for peaceful use. Since that time, many space treaties and laws have been established. Briefly, the international agreements regulate activities in space in the following manner:

- No military bases, maneuvers or weapons tests will be conducted on Mars or other celestial bodies.

- Systems placed in space or on celestial bodies remain the national property of the owning government. Such ownership will not be affected by the system's presence in space, celestial bodies, or its return to earth.

- Governments will register with the Secretary General of the United Nations as soon as practical after launching of their space objects (Convention on Registration, 1971).

- Outer space and celestial bodies cannot be claimed as sovereign territory by any nation.

- No harmful contamination of any celestial body.

- No use of environmental modification techniques to alter, destroy, or damage another state.

The above points are not all inclusive but do point out the eagerness to preserve the national habitat of space. However, international laws do permit the use of military functions such as communications, navigation, meteorology, reconnaissance, and surveillance. They also permit the use of military space stations, testing, and deploying nonnuclear and non-ABM weapons systems for self-defense.

1.3 U.S. SPACE STRATEGY

It is important to point out that the United States does not have a clearly defined space strategy. A strategy is quite different from a policy or law. The Congress has the Senate Committee on Aeronautical and Space Sciences which deals with legislation on space activities (except for those peculiar to DOD), matters relating to the scientific aspects of space activities (exclusive of military) and matters relating to NASA. The House has the Committee on Science and Astronautics and deals with legislation on space technology, international agreements and activities of NASA, manned space flight, science, and research and development. The strategy of the United States is governed by the policy goals of each federal administration. For example, President Bush's administrative space strategy was to:

- Strengthen the security of the United States

- Maintain United States space leadership

- Obtain economic and scientific benefits through the exploitation of space.

In order for the U.S. space strategy to grow, it must evolve from a joint effort of the administration and the Congress. President Reagan's national space strategy was based on a White House fact sheet for public release on 7 September 1984. He cited three programs: i) the Civil Space Program, ii) the Commercial Space Program, and iii) the National Security Space

Program. As we are interested in the unmanned planetary studies, we need point out the key features of only the Civil Space Program:

- Establish a permanently manned presence in space. The goal was to achieve this by 1995. (Due to the Challenger disaster, this goal was not met).

- Foster increased international cooperation in civil space activities. (This goal was partially achieved by a joint Soviet-U.S. program to study for a possible future effort to land on Mars.)

- Identify major long-range national goals for the civil space program. (This was seriously affected by the Challenger disaster as well as the successful planetary missions.)

- Insure a vigorous and balanced program of civil scientific research and exploration in space. (This has been mitigated due to huge budget cuts in the space programs.)

There is no question that domination of space could result in control of considerable lower altitude activities. In the recent past, the U.S. space strategy appeared to be coalescing in favor of military exploitation of space. For the present, the U.S. space strategy is to gain more knowledge about earth, its neighbors, and to aid in the comfort and well being of its inhabitants.

1.4 THE MARS PROJECTS

As we pointed out earlier, NASA has two prime functions: to serve the military and to advance science. In the latter function, NASA supplied the means for scientists to study our Solar System. The Mariner program was one of the most ambitious scientific explorations ever performed. Mars exploration began on 15 July 1965 when the unmanned Mariner 4 spacecraft flew to within 9,780 km of its surface. The pictures from the spacecraft revealed a harsh, lifeless, heavily cratered Moon-like surface with no evidence of canals, vegetation, or seas. Mariner spacecrafts 6 and 7 were launched four years later in 1969 and revealed a Martian carbon dioxide atmosphere at very low pressures and temperatures. Mariner 9 was launched from Kennedy Space Flight Center on 30 May 1971, and on 14 November at 00:15:29 GMT it became the first spacecraft to orbit another planet. The objective of this mission was to explore Mars from orbit for a period of time necessary to observe most of its surface and certain specific areas for dynamic changes. Measurements of both the atmosphere and surface were made.

The Mariner 9 project made startling discoveries. A volcano the size of the state of Arizona and 15 miles high (almost three times the height of Mt. Everest) was discovered. Also discovered was a gigantic channel over ten times the size of the U.S. Grand Canyon, evidence of both liquid and wind erosions, some catastrophic flooding, and enormous river beds. Both high and low resolution cameras provided means whereby most of Mars could be photographed, as well as detailed enlargements of particular interest to scientists. Some of the Mariner 9 photographs are contained in this book.

The second Martian program was Viking. It was started in the late 1960s before any results from Mariner 9 came in. Its mission was to detect life on Mars by using a landing vehicle to perform some laboratory experiments. Two Viking spacecraft were launched in the summer of 1975 and went into orbit around Mars approximately a year later. On 20 July 1976 one of the landers made the first soft landing on Chryse Planitia, 22.5°N, 48.0°W. The second spacecraft arrived two and a half weeks later on 7 August and landed on 3 September at 44°N, 226°W. There followed a number of experiments. The biology experiments searched for life by examining the soil chemistry. Metabolism was simulated by adding organic compounds to the soil, allowing it to decompose. Photosynthesis was simulated by adding water, causing release of oxygen. A gas chromatograph-mass spectrometer recorded no organic molecules in the soils. Inorganic analysis was performed by an x-ray fluorescence spectrometer. Marsquakes were not detected by a three-axis seismometer. Atmospheric properties measured by sensors, a mass spectrometer, a potential analyzer, and mapping of the atmospheric water vapor and surface thermal properties by infrared instruments were performed. (These were the main objectives of the Mars Atmospheric Water Detection (MAWD) and the Infrared Thermal Mapper (IRTM)).

The landers performed way beyond their expectations. The second lander ceased operating approximately four years after transmitting data, and the first lander more than doubled that. The orbiters observed clouds, storms, and the chemical composition of the atmosphere, as well as fluctuations in the behavior of the polar caps and distribution of surface debris for nearly two Martian years. The first orbiter was lost in August 1980 and the second in July 1978, a remarkable testimony of the ingenuity of man.

2

THE BUILDING BLOCKS
FOR LIFE ON MARS

When we inquire into the possibility of life on Mars, we must ask what are the essential building blocks required for life to form. Most scientists would probably concur with the following basic requirements.[3]

2.1 CHEMISTRY BLOCK

Most everyone will agree that carbon is the sole basis for the "spontaneous" generation of life, that nitrogen, oxygen, and hydrogen are also required, and that these elements must be in appropriate quantity.

2.2 WATER BLOCK

To imagine life without water is solely an academic exercise. Water is an absolute requisite for chemical reactions in cell formation, in cellular transportation, and cellular interactions. Water is one of the most plentiful of fluids, and yet most elusive to describe mathematically. Earth's capacity contains 1.5 billion cubic kilometers of water in one form or another. Its extraordinary physical properties endow it with a unique chemistry.

Let us examine[4] a few of its macroscopic properties, and then we will discuss its structure. Like everything else (ice included), liquid water contracts when it is cooled, but the shrinkage ceases before solidification, at about 4° Celsius. For temperatures lower than that, the cooler water lies on top of the warmer. Ice has a specific gravity of 0.92, and thus an ice floe will float in water with about an eleventh of its volume above the water surface.

Water has the greatest specific heat known among liquids. Water's latent heat of vaporization at 20°C requires 585 calories to evaporate just one gram of water. With the exception of mercury, water has the greatest thermal conductivity of all liquids. Hence, water is a fantastic energizer of our atmosphere.

In basic structure, the water molecule has a small dipole moment and is barely ionized. Water will dissolve most anything in time. The dissolved material tends to remain in solution due to exceptional attributes. The values given by the inverse square law for the force that attracts separated positive and negative ions are determined by multiplying the square of the distance separating the ions by a constant that varies according to the nature of the separating medium. Known as the dielectric constant, water possesses the largest value compared to any other substance. Thus, due to this dielectric constant, water is not pure (unlike water vapor which can be pure). So water can pick up hydrogen ions, and thus its chemical structure is quite variable.

Another way to describe water, besides using hydrogen ions, is to identify its free energy (pF). Material in solution, whether it is ionized or not, disturbs the liquid structure of water; in thermodynamic terms, the presence of solutes decreases the pF of water.

The quantity measured in pF units is a potential with dimensions of pressure. The pF of water can be decreased, for instance, in capillary systems. The energy to lift water into a capillary tube (or into plants or rock fissures) comes out of the pF of water. The value of the potential is found by doubling the known value of the liquid's surface tension and dividing the product by the liquid's radius of curvature. Water has the greatest surface tension of any known liquid.

Liquid water absorbs heat radiation so intensely that if our eyes were sensitive to the infrared, water would appear black. The fact that water absorbs red light accounts for its characteristic blue-green color.

One of the surprises of the century was that water is not simply H_2O, nor is it a single substance. Harold Urey discovered that the purest water contains deuterium. Lately, we find a third isotope of hydrogen, called tritium, and three isotopes of oxygen: 0-16, 0-17, and 0-18. Hence the purest water that can be prepared in the laboratory is made of six isotopes, which can be combined in 18 different ways. So when we add the various ions we find that pure water contains no fewer than 33 substances.

Let's go back to Urey's deuterium. For instance, D_2O has a higher boiling point than H_2O, is more viscous than tap water, and nothing grows in it. Animals die of thirst drinking it.

We know life, as we understand it on earth, cannot exist without water. Our atmosphere requires it, and the protein molecule (the basic material of all life) depends upon it.

What of its structure? Let us examine the structure of ice. The two hydrogen atoms are bonded to the oxygen atom approximately at 105° to each other. Now if the angle between the centers were say 110°, then the frozen water molecule would form a cubic lattice (as in the diamond crystal), and the structure would be unstable due to the strain on the distorted bonds. The precise arrangements of the molecules in an ice crystal are not known. We know they form a hexagon and that the molecules are joined by hydrogen bonds.

The forces of attraction between molecules in ice or water produce strong inward pressures. If one thinks of the open structure of the ice molecule's arrangement as the arch of a bridge with strong downward stresses, then when the ice temperature increases, the thermal agitation of the molecules causes the ice structure to collapse, and the substance is now a liquid. It is known that application of pressures from outside the ice will make ice melt at a lower temperature, and this reinforces the internal pressure within the ice and assists in its collapsing.

The chaos of the water molecules is wild. The angle between the two hydrogen molecules is no longer constant. Each oxygen atom now attracts, by electrical forces, not just two extra hydrogen atoms as in ice, but there is more. Thus the hydrogen atoms are continually shifting, and it is this random displacement that affects the viscosity and dielectric constant of water.

2.3 EXTRATERRESTRIAL BLOCK

When we consider our universe as a single entity, we are addressing stars that number 100 billion billion, or viewed slightly differently, there are 1 billion stars in each of the 100 billion galaxies. Millions of these stars are millions of time brighter than our sun. When we consider life forms on other planets such as Mars, we must consider not only the appropriate size and mass, but the appropriate distance from the sun. And the planet must have specific materials to promote and sustain life. We shall discuss each of these later, but first we must discuss the evolution of the universe in order to set the stage to discuss life in the universe. Then we can discuss the evolution of the earth to set the requirements that scientists see as the requirements for life on Mars.

Much has been written on the evolution of the universe, Riordan and Schramm's book[5] being the source of much of what is to follow. The precise time when the universe began is one of the great questions in cosmology. Part of the mystery is due to a total ignorance of its source and how it came to be, from whence came the mass of the universe, the potential and internal energies, the structure and the cause and trigger that initiated the universe. According to some of the more popular cosmological theories, the universe started approximately 15 billion years ago in a region much, much smaller than a volume of a pea. Something triggered a massive "explosion," if that is the correct word to describe the beginning, having initial conditions that are presently beyond understanding. At that instant, the temperature was supposedly "infinitely" hot, and the density "infinitely" large, "infinitely" being the word to describe a magnitude so large it cannot be defined; yet, truly not infinite in the mathematical sense as something infinite never becoming finite.

There are three reasons given for believing this was the manner in which the universe began. The first reason is that we observe space to be heavily populated with galaxies that formed approximately 1 billion years ago, and these galaxies are moving apart from each other, which is exactly what should occur if a Big Bang occurred. The second reason is that the universe has a uniform distribution of electromagnetic radiation at a near constant radiation temperature of 3 degrees above absolute zero, which is exactly what is predicted theoretically. The third reason is that there exists certain light elements ^{4}He, ^{2}H, ^{3}He, and ^{7}Li, which do not originate from the formation of stars and galaxies but must have come from the beginning. It is significant to point out 25% of all matter in the universe is ^{4}He, and 74% is ^{1}H.

It has been postulated that between the Big Bang and Planck's time (10^{-43} seconds), the physics of the universe is undefined. We have a theory proposed by Hawking and Penrose that the universe started from a singularity that likely was at the center of an enormous black hole. In relativity, there is no way to create an expanding universe other than from a singularity.

In calculating the energy content of the universe E, we consider the famous equation

$$E = 4\pi R^3 \rho c^2/3 = c^5 t/8G \tag{1}$$

where R is the spherical radius of the universe, ρ its density, c the velocity of light, and G Newton's gravitational constant. The impreciseness of Eq. (1) is $\hbar/t$, where $\hbar$ is Planck's constant. One solves for t in Eq. (1) yielding the result $t = 10^{-43}$ s. Hence, for $t < 10^{-43}$ s, we do not know what to say (uncertainty), and for $t > 10^{-43}$ s we are more confident in what took place.

Using the above equation, we find the radius of the universe at Planck's time $R = ct_p = 10^{-33}$ cm, a distance so small it is difficult to envision it. We can also calculate the mass of the universe using $m_p = \hbar/tc^2 \simeq$ 10 micrograms. Hawking views the universe prior to Planck's time as "a seething foam of forming and evaporating black holes," because black holes prior to 10^{-43} s yield a mass < 10 μg that would evaporate.

During those very early periods after the Big Bang, a theory was needed to explain the drop in temperature from "infinite" to several millions of degrees (a drastic cooling). To allow for this dramatic cooling, Guth proposed an inflationary model that could bring about a rapid exponential expansion which lasted 10^{-12} s. In this soup the fundamental quantum particles of chemistry formed, and the formation of protons, neutrons, and other hadrons took place up to 10^{-5} s after the Big Bang. The soup contin-ued to foment up to 10^2 s, which marked the end of nucleosynthesis. The creation of new particles slowed down due to the cooling and expansion of the universe. Almost all the particles and antiparticles that were being created and annihilated disappeared. Fortunately, there was an excess of electrons over antielectrons, and quarks over antiquarks to provide for those particles existing today. This very early excess of matter over antimatter (about one part in 10^{10}; one out of ten billion) that survived gave us the atomic nuclei that after a million years later formed the atoms that later were cooled to heavier elements in stars that eventually provided the elements needed to have life.

There were small inhomogeneities that existed when the universe was 10^{-12} s old that triggered the formation of the first stars and collections of stars which we call galaxies. It is interesting to note that we cannot observe any inhomogeneities prior to a million years after the Big Bang, which was when the universe first became transparent.

The Standard Model posits an electromagnetic field whose quanta are photons; an electric field whose quanta are electrons and antielectrons; and other fields whose quanta are leptons and antileptons, quarks and antiquarks; plus fields whose quanta are particles that transmit the weak and strong forces that are so necessary in the action of elementary particles which govern our universe on a microscopic scale. The Standard Model is a workable law of physics but is not the final law of nature. There is much about this law we do not know. It does not include gravitation, for example. We already have a marvelous field theory of gravitation (General Theory of Relativity), however the quantized version of this theory breaks down at high energies.

There is a new theory (String Theory) where the quantum field theory of the Standard Model is transformed in string theory as being tiny one-dimensional strings that can vibrate and behave in motion just like the classical interpretation. The use of the word string refers to the fact that the particles are not points (electrons, quarks, etc.) but are particles about 10^{-33} cm long that can have two open ends or can be fashioned into a loop. Superstrings are quantum theories based on one-dimensionality. They have remarkable properties that satisfy many of the perplexing problems of many other theories. For one example, superstrings incorporate the equations of general relativity in the limit. Two of the great appeals of superstrings are that superstrings require the existence of gravity and that they function in many dimensions. Thus gravitation can be interpreted by string theory. Though superstrings are very promising, they too have many unanswered questions.

It is profoundly amazing to contemplate the fact that life as we know and understand it would be impossible if any of the constituents of the aforementioned building blocks of life had slightly different values. Such as changing water H_2O (which is so essential for life) to H_2O_2 (which cannot sustain life). Or consider the energy of just one of the excited states of the carbon 12 nucleus. When two helium nuclei join to create the nucleus of beryllium 8, which sometimes absorbs another helium nucleus, fission forms the excited state of carbon 12. This nucleus emits a photon, decaying into the lowest energy state. In the subsequent nuclear reaction, carbon is built up into oxygen and nitrogen plus other elements so essential for life. This capture of helium by beryllium 8 is a simple resonant process. If the excited state of carbon 12's energy were just slightly greater, the formation rate would be less, resulting in the beryllium 8 nuclei fissioning into helium nuclei before any carbon could be formed. Thus, the universe would contain almost solely hydrogen and helium, with none of the elements for life. Just a slight change would result in disastrous results for life.

Another example of a slight alteration concerns the vacuum energy necessary to explain the rate of cosmic expansion. If the various contributions to vacuum energy did not approximately cancel out, the universe would have gone through a cycle of first expansion (our present conditions) and then contraction back to the initial singularity before life could form—or keep expanding that no stars could have formed. The total vacuum energy must be just small enough (accurate to about 120 decimal places) for life to be possible. But enough of these "happenstances." Let us return to the building blocks for life.

2.4 PLANETISMAL ELEMENTS BLOCK

The elements that comprise our earth and all life forms are the result of the explosions of ancient suns that were formed billions of years ago from the basic elements of hydrogen, helium, and beryllium which we discussed earlier. The nuclear reactions within the stars produce the elements from which all living things are made. Fortunately, that nuclear fusion process

occurred *steadily* for billions of year allowing time for life in its most elemental form to develop into intelligent life.

A star is a spherical mass of gas that is delicately balanced between two
forces: a gravitational attraction force and an outward pressure force due
to the hot gases within. The compressed gas of hydrogen has the approximate density of water (10^{36} times greater than the norm). Its internal temperature may reach a temperature of 15 million Kelvin. At this temperature, electrons disappear from the hydrogen atoms. The hydrogen nuclei
which are protons fuse into helium nuclei (two protons and two neutrons)
emitting two neutrinos, two positrons, and enormous energy. In time, the
star's core shrinks and temperature rises to maintain the pressure balance.
The star loses its homogeneity. The core gets smaller, whereas the outer
layer enlarges up to 50 times its former radius. This phenomenon results
in the star (the size of our sun) transforming into a luminous red giant. If
it occurred to our sun, earth would be vaporized into gases.

Inside the red giant, the core contracts, the density and temperature increase. The helium that was formerly created from the star's hydrogen
now becomes the star's fuel. The collision of two helium nuclei produces
beryllium which, as we explained earlier, now leads to the production of
carbon. Oxygen is formed by fusing one more helium with carbon, and
now we have one more building block for life: carbon and oxygen.

The red giant exists only for a few hundred million years. Its last death
throes of combustion are quite unstable. The red giant discards its outer
mantle revealing a shell of gas named a planetary nebula. Matter enriched
in carbon is convected up from the giant's core and is outgased to space as
graphite. Soon the helium is exhausted, the core solidifies, and the red
giant transforms to a white dwarf. In time it cools and can disappear from
detection.

A different scenario exists for a star larger than our sun. A star of, say,
20 solar masses is correspondingly 20,000 times brighter than our sun. It
becomes a red giant 1/1000 the time for our sun to become a red giant.
Due to its massive size and temperature a different chemistry exists. The
carbon nuclei fuses to make magnesium, the oxygen fuses to make silicon
and sulfur. The silicon in turn fuses to make iron.

Once the star has an iron core, the core no longer generates energy
through fusion. The core collapses inward in only one second and becomes so compact that no material can enter it. Thus, material that is falling in towards the core bounces back towards the surface creating a pressure shock wave. When the shock wave reaches the star's surface, it ex-
plodes, creating a luminosity as bright as a billion suns. The energy of
such an event is greater than all the energy expanded in the life of the star.
Such is a super nova.

A super nova ejects its chemical materials: helium, oxygen, carbon, sulfur, iron, and silicon. Behind the shock wave, intense heat creates nuclear products that are radioactive (uranium), and stable elements like gold and lead. This then is a rough explanation of the origin of elements that are found in the interstellar material of galaxies.

In our Milky Way galaxy, there are vast regions of spiral arms that are dust which shield the hydrogen atoms from ultraviolet, enabling the hydrogen atoms to form H_2 molecules that can combine with oxygen atoms creating water. In addition, ammonia NH_3 is formed. Just recently, Miao and Khan found glycene (one of the smallest amino acids) in a gas cloud near the center of our galaxy. These materials could easily be attracted and attached to meteorite that could impact on such cold and forbidden planets like Mars and earth. There is very little dispute among scientists that our solar system consists of elements formed by spent stars billions of years ago. Thus, the material for future suns and planets is locked up in the furnaces of the billions of stars in our universe.

Stars are formed by coalescing the gases in space into rotating disks and hence into spherical masses. Some of the gases coalesce into giant spherical masses, like Jupiter, and other combine with the dust, compress, and form planets like earth and Mars.

2.5 PLANET'S ATMOSPHERE BLOCK

Just as gases permeate our universe, so do gases that attach to planets. Before we discuss the origin of atmospheres, let us consider earth's atmosphere, which we know supports life. At earth's surface, air is mostly nitrogen, about 21% oxygen, plus small amounts of carbon dioxide, xenon, neon, argon, and krypton. Mixed in this air is water vapor and particulates consisting of meteorite dust, industrial wastes, and volcanic ash. The atmosphere is popularly considered to be 400,000 feet thick, though air has been detected above 600 miles.

At nine miles above the earth's surface, photochemical processes take place, the most significant being solar radiation of ozone, a form of oxygen. At 14 miles, the layer of ozone is maximum. Most everyone is aware of the importance of this ozone layer as being the principal absorber of solar ultraviolet radiation. At extreme altitudes, one finds such oxides as nitrogen oxide and deuterium oxide. No estimate is known of the gravitational de-mixing of the atmosphere, except that solar and cosmic radiation occur which radically change the molecular properties of the various gases.

We are aware that some of the atmosphere of Mars has degassed to space. Degassing is due to two factors: the density and temperature of the atmosphere at lower altitudes, and the gravitational pull of the planet. To escape earth, matter must exceed seven miles a second, but half that (3.1 miles per second) to escape Mars. The greater the escape velocities, the lighter will be the remaining gases since light molecules have greater velocities than heavy molecules. The speed of individual gas molecules depends on both temperature and density as well as the direction of molecular impact on other molecules. For example, it has been estimated by Landsberg[6] that it would take 10^{45} years for nitrogen and 10^{51} years for oxygen to disappear from earth. Compared to the earth's present age of 4×10^9 years, these gases will remain on earth for some time to come. Helium, on the other hand, is continually being degassed to space. Helium is constantly being formed on earth by radioactive processes, yet no accumulation has ever been detected.

So how did our atmosphere form? First, it is assumed that earth was formed at a time the other planets were forming, and thus the elements of the planets which have been discussed earlier are similar. We know the planets' temperatures were much greater than they currently are. For example, at 1600°C, almost all gases in the host atmosphere would degas to space. As the temperature decreased, water vapor, nitrogen, and carbon dioxide, substances found in contemporary volcanic gases, formed from the cooling of molten rock. Cooling further, the water vapor condensed and oceans formed. Simultaneously, oxygen in small quantities would have been liberated from the breakdown of water. Since oxygen is the second most abundant element on earth (nitrogen being the most abundant), we must ask where oxygen was derived. Landsberg proposes three possibilities.

Certainly, oxygen can be formed by dissociation from water. However, oxygen would immediately recombine when it came into contact with the very hot rocks. A second possibility is again dissociation from water vapor at high altitudes, but it is difficult to envision enough water vapor existing at these higher elevations to form the enormous quantity of oxygen needed even over a period of 4 billion years. Certainly some of the oxygen is formed this way. The last hypothesis is oxygen being created by photosynthesis of carbon dioxide in plants emitting oxygen. The process has some scientific basis. It is postulated that life could have begun in the oceans if the water temperature was 61°C or lower. The purple sulfur bacteria produces organic matter in a strictly CO_2 environment. Such bacteria could be the earliest forms of life and the precursor of green plants that produce oxygen.

Mars' atmosphere has a lower density than earth's, and thus it has a much lower escape velocity. Spectrographic analysis has revealed Mars' atmosphere consists of nitrogen, carbon dioxide, water vapor, and argon, but no oxygen. Whatever oxygen might have existed could have out-gassed to space or oxidized with the minerals. So how did Mars acquire a climate so different from earth's when both had similar elements and both were temperate to maintain liquid water? The answer is that the two planets differed in how they transported carbon dioxide between their surfaces and atmosphere. Earth has maintained its moderate climate due to its ability to increase carbon dioxide to the atmosphere when the surface cools and reduce carbon dioxide when the surface temperature rises. Mars has lost the ability to transport carbon dioxide back to its atmosphere. Key to a planet's evolution, especially its ability for life, is carbon dioxide's role. Carbon dioxide acts like a sponge. As the planet's temperature rises, carbon dioxide in the atmosphere declines, thereby cooling the surface. When the surface cools, carbon dioxide increases, thereby warming the surface. This delicate cyclic behavior permits sunlight to either enter or not enter the lower level of the atmosphere, hence acting as a thermostat for the planet's surface temperature.

2.6 THE CARBON-DIOXIDE CYCLE BLOCK

A mathematical model[7] was developed that helps explain the current status of the Martian climate. It is highly probable that once long ago Mars had an adequate amount of carbon dioxide in its atmosphere, probably enough to have kept a large part of its surface (especially around the equator) from freezing. But for some as yet unexplained reason, the cycling process mentioned earlier failed. But before it failed, it was sufficient. If this hadn't been the case, the weathering of the Martian rocks would certainly have removed the atmospheric carbon dioxide within the short space of only 10 million years. Examination of the vast network of channels (to be discussed shortly) show the networks existed close to 3.8 billion years ago.

The recycling of carbon dioxide is assumed responsible for removing carbon dioxide from the atmosphere by the same geophysical processes we observe on earth. Mars probably cooled down because it was too small, had less internal heat than earth, and lost more heat at a greater rate than earth. This heat conduction lowered the planet's internal temperature resulting in Mars being so cold it could no longer release carbon dioxide from its carbonate rocks. That, plus the likelihood that any carbon dioxide which left the Martian atmosphere was absorbed by the minerals and remained locked in the crust. The atmosphere got thinner and thinner until it reached its present state. All Mars needed was greater mass so it could have the internal heat to recycle the carbon dioxide, thereby counteracting the insufficient amount of sunlight it was receiving.

To prove this mathematical mode, one needs to find the buried carbonate rocks in Mars' crust. The current NASA Mars space program should resolve this.

We know, for example, that sea floor spreading can release CO_2 to the atmosphere resulting in atmospheric temperatures rising. We know that a low quantity of CO_2 in the atmosphere can produce glaciation on earth. By examining the uplift history of the mountains on Mars, we can grasp what the climate of Mars once was. For example, by studying strontium 86 (which is in volcanic emissions) we can tell whether mountain erosion or mountain washing into the oceans from chemical weathering occurred. We should find by retrieving rocks near Martian uplifts whether there exist fossilized plants in certain rocks, indicating rivers that existed millions of year ago. One would need a wide range of different plant fossil to determine events with any certainty.

Certainly the picture is quite unclear as to what exactly took place on Mars. Many scientists will say all that has been presented herein is too simple; however, a great many concur with what has been stated. Regardless, the mystery as to exactly what transpired may never be resolved until tests and theories agree, and for that to occur vast explorations on this mysterious planet will have to be conducted.

2.7 RIVERS BLOCK

We shall soon be discussing the many channels on Mars. Many scientists believe these channels were formed by catastrophic flooding. Before we postulate our analysis, we first need to know how rivers are formed.

A river is simply a system of drainage as a result of liquid moving from a greater potential (elevation) to a lower potential by gravitational means. Even in flooding, a river remains within valleys through which channels run.

Erosion is the mechanism through which water flows naturally. Water carves channels, keeping them somewhat free by the abrasive action of the sediment carried by the flowing water against more solid forms impeding the water's progress.

There are many records showing the flow of massive quantities of water over areas not normally prepared to experience it. These catastrophic phenomena can be due to the spillage from a dam, the rupture of an ice wall or glacier field, or the impact of a meteor into a large body of water. Two regions on earth point to catastrophic flooding. The first is the Channeled Scabland on the Columbia Plateau (see Figure 1) in the state of Washington.   The central portion of the photo shows the channeled

Scabland composed of strongly dissected terrain consisting of canyons and outliers in basalts overlain by loess. This topography was formed by massive flooding, probably after a larger glacial lake rapidly emptied following failure of a "natural" dam composed of terminal moraines laid down at the edge of a continental glacial ice mass. Though this area is now dry, satellite photo-graphs of the region show an impressive system of channels that at one time drained more than 10,000 square miles of wetland.

A second region of catastrophic flooding is the Caroni River in Venezuela. The Caroni River is a most interesting example of a vast system of drainage channels.

The Caroni River is not truly a river. It possesses no trunk channel. By definition, a river is a system of drainage from tributary channels joining a trunk channel—like a local stream joins a larger stream (a river) which then empties out to a bay. The bay in this instance is a trunk channel. Both the Caroni and the Scabland have similarities.

- both have extensive gravel deposited across the drainage area

- both have randomly oriented slopes inclined toward undrained depressions

- both regions have bedrock uplands that include drainage divides.

Based on the studies of erosion in these two remarkable examples of earthen catastrophic flooding, a clear understanding of what might have caused the drainage patterns appears. The major cause was a dramatic change of climate—from arid conditions to humid ones. As pointed out by H.F. Garner, if the environment is nonglacial, i.e. there is no freezing such as in the Caroni flows, only a breakdown of rocks and boulders that one finds in deserts could explain the coarse gravel. This breakdown is probably due to extremes of heating and cold producing internal stresses that crack the rock. To transport this sediment requires massive flooding. The waters, saturated with sediment of all sizes, will flow from wide channels of deep depth to numerous smaller channels of shallower depth. The deeper the depth, the more lasting the channel.

Water flowing in multiple channels in various systems (or networks) will flow along paths of least resistance. This resistance can be due to increases in elevation, and it can be due to surface resistance like divides, rocky mesas, and enormous boulders. Examination of the behavior of water flowing in shallow channels existing in a field or plateau show that the network of channels can recombine to form a trunk channel, very much like a network of parallel pipes or electrical circuits forming a trunk line.

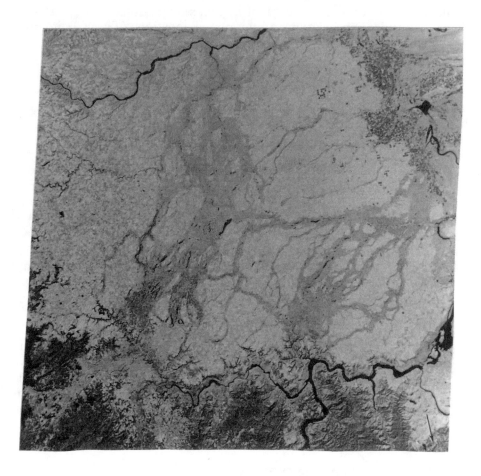

Figure 1. Satellite photograph of channeled scablands of eastern Washington. Courtesy of NASA.

Changes of climate appear to be a mechanism that establishes conditions ripe to both form rivers and feed the rivers. One mechanism is a general warming of the oceans and lakes. Air rises from warm zones to cooler ones, moving from the seas inland. The air carries the moisture from evaporation of surface water and falls as rain at cooler temperatures existing at higher altitudes. Obviously, the distance that water vapor is carried is dependent on the temperatures of the seas and air. Thus, by examining the carbon dioxide content in rocks, elevations of huge areas of plateau, altitudes and locations of mountains, locations of seas, latitudes of the planet and air currents, and temperatures and densities, a mathematical model can be devised that predicts with excellent accuracy rain forests and zones of greater than normal precipitation. For example, the great Tibetan plateau adjacent to the Himalayas provides the morphology for the condensation of the massive water vapor from the Middle East (Persian Gulf) that transforms into the monsoons in the Far East.

The mechanism described above operates in regions about 35 degrees north or south of the Equator, which is roughly two-thirds of a planet's surface. The colder the sea, the farther the air has to rise to produce rain. Hence, when the oceans are cold (as in periods of glaciation), rain may miss the coastlands entirely, falling on high ground deep inside the landmass. And the opposite occurs. When glaciation subsides, oceans grow warmer, then rain can occur closer to the coast. This seems to be the case with both the Scabland of Washington and the Caroni River, the former now arid, the latter wet. Other dried up river networks that carried massive quantities of water are ancient. The Chad district of central Sahara between Lake Chad and Tibesti is an ancient network of river channels.

We shall show similar ancient networks of channels on Mars that suggest the strong likelihood of flooding. With water and carbon dioxide, we now have the basic constituents for oxygen. If at approximately ±35 degrees latitude the temperature of the water on Mars was 61°C, then one has the setting for life to exist. But first, we must discuss how water freezes to learn if such conditions could exist on Mars.

2.8 FREEZING WATER BLOCK

One of the major issues in discussing life on Mars is not only whether there is or even was water on Mars, but the properties of water if water does not exist. We shall discuss the pressures versus temperatures of liquid water and ice in Chapter 5, but first we need to discuss how water freezes on earth, and then how we expect it freezes on Mars.

The equilibrium temperature at which water freezes on the earth's surface is 0°C. The equilibrium temperature varies with pressure. For example, at one atmosphere (14.7 psia), the equilibrium temperature (e.t.) is

0.0072°C. Each temperature and pressure of an ice crystal depends upon the critical radius of surface curvature. The critical radius of curvature determines whether the ice crystal will melt or freeze. Hence, pressure, temperature, and geometry are the factors that determine how water freezes.

For equal amounts of liquid water and ice mixed together, the temperature of the mixture will reach e.t. since heat given off by water is absorbed by ice. We call this heat exchange the latent heat of fusion. It is the heat exchange that occurs as a substance melts or freezes at the e.t. It is precisely the energy per pound mass (lbm) required to alter the highly organized solid crystal structure of an ice crystal into the random continually changing behavior of the molecules in the liquid phase. On earth's surface, the latent heat of fusion is 80 calories per cubic centimeter. The rates of freezing and melting depend upon both temperature, pressure, and surface curvature. At a given temperature, there is only one pressure when the two rates are equal. At a higher temperature, the melting process dominates, and at a lower temperature the freezing dominates.

For water that is not pure but has minerals such as salt dissolved in it, the freezing point is lowered. The minerals slow down the freezing process by decreasing the number of molecules that are available. Thus, the freezing point is lowered.

Liquid water cannot be cooled below the freezing point at a fixed pressure. Removing any additional heat from the ice simply causes an increase in the volume of ice, not a lowering of temperature. But if there is no ice in the presence of liquid water, then the liquid water can be "supercooled;" that is, the water can have a temperature below its freezing temperature. However, if ice is dropped into this supercooled liquid, the liquid immediately freezes.

The temperature of equilibrium strongly depends on the curvature of the free-surface. That is, the e.t. is lower for a small ice sphere than a large ice sphere. Figure 2 is a plot of the critical radius versus the freezing temperature. The figure shows that liquid water can be supercooled to temperatures below 0°C providing there are no particles in the water with a geometry having a radius of curvature larger than the critical radius for that temperature. Note, at 0°C the critical radius is infinite, i.e. no curvature. This is the case of flat surfaces. So much for the basics. Additional background is found in standard thermodynamic texts.

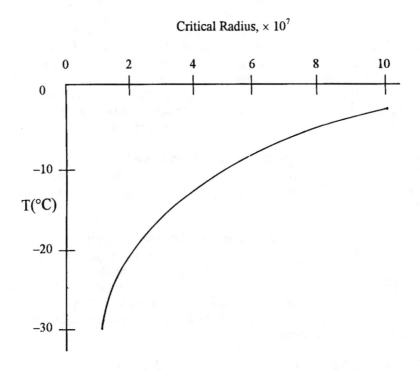

Critical Radius, $\times\ 10^7$

Figure 2. Freezing Temperature of Water Versus Ice Crystal Shape

Let us now focus on what may have occurred on Mars. Following are arguments proposed by Bruce Chalmers[8] that stemmed from researching the solidification of molten metals. The heat of fusion has to be transferred from a region where ice is forming if freezing is to take place. What customarily happens in the freezing of a lake is that heat is transferred by conduction through ice that is already formed. The interface between the ice and water is flat since any region that has a convex curvature is cooled with less efficiency than the rest of the region. That is why there sometimes is an annular ring of liquid water around a rock in a bird-bath where birds can still drink despite the majority of the surface water being frozen. In supercooled water, the latent heat of fusion is a transfer of heat outward to the supercooled liquid, where cooling is more efficient at convex regions. So convex regions become more convex. This heat transfer, produced by the increase of convexity, suppresses growth in the region resulting in a branching or "dendritic" growth.

Such dendritic growths as a result of this supercooling process are offered as a possible explanation for some of the strange surface phenomenon on Mars. The dendritic patterns will be discussed in Chapter 4.

3

SETTING THE STAGE

When one uses the title *Life on Mars* as the subject of a monograph, the reader conjures up many topics that are as varied as can conceivably be imagined. The subject is enormously complicated, and the answers to such questions as where did the water come from, where did it go, and what mechanism carved the great channels on Mars, will never be answered satisfactorily until a thorough examination of the planet takes place.

The literature on the subject is vast. A number of papers are classic. One of the best papers is that of Squyres [1]. But this Urey prize-winning sixty-page article addresses only two aspects of the problem: the possible existence of standing bodies of liquid water very early in Martian history and the present distribution of Martian ground ice. Many other important topics dealing with water on Mars exist such as i) what is the hydrological cycle on Mars (we suspect it differs from earth), ii) what is the precise mineralogy of the abundant Martian dust, iii) what is the dust/ice ratio of the polar layered deposits, iv) what is the porosity of the Martian brecci-ated material and how thick is it, v) what is the rheologic behavior of ice on Mars, vi) what was the effect of climate on the water distribution, no-tably the effect of CO_2 mass flux, vii) what are the mechanics of Martian thermokarst, and viii) what is the precise history of ground water sapping channels? These and other questions may never be solved until man per-sonally explores Mars.

The purpose of this book is to address a few of the above questions and summarize what others have determined, point out the areas of

26

controversy on a few issues, and pose some areas of research that could unlock some of the mysteries dealing with water on Mars and therefore, the likelihood for life. To accomplish all this, we shall first present the current assessment of the evolution of water on Mars, the evidence for water both geomorphically and chemically, its loss and storage, and the cycles in climate and transportation of water. We will then present some of the models and attempt to assess their importance.

Much of what is to be presented is based on intense analysis by NASA scientists, both government and private geologists, and on vast data from the Viking and Mariner 9 missions. The data have shown conclusively that water, both in liquid and solid states, has played a key role in the Martian morphology. There is no question that vast quantities of water once existed on Mars and may still exist, lying below the surface. There is chemical evidence that water has been outgassed, but the precise mechanism driven by climatic changes is not firmly established due to a lack of material evidence.

Before we become involved in the interpretation of the data from the Viking lander and Mariner 9 space vehicles, consider the following possible scenario. About 4.6 billion years ago, Mars experienced catastrophic outgassing equivalent to much less than 1 km of water spread over the en-tire surface of Mars. As the dense water-rich atmosphere cooled, an equilibrium between liquid and vapor phases occurred. The liquid phase created the small valley networks and percolated through the highly porous surface to develop into immense ground water systems. As the atmosphere continued to cool, this liquid ground water was trapped, never to evaporate into the atmosphere due to subsurface freezing. The atmosphere evolved for billions of years, volatiles being added by volcanic eruption, much of the nitrogen and water in the later atmosphere being lost by exospheric escape. The great reservoir of ground ice was now trapped due to extremely low temperatures. Yet various processes temporarily released melted ground ice or confined water beneath that ice to the Martian surface causing catastrophic flooding resulting in gigantic channels being carved through the terrain at specific latitudes. This is an interesting but speculative scenario, one that might be kept in mind as we preview what is known and what is suspected about the water on Mars.

Before we make our presentation, it is significant to make a comparison of atmospheric constituents on earth and Mars. This will be a nice reference to refer to as topics are discussed.

TABLE 1. Comparison of Properties and Atmospheric Constituents on Earth and Mars

Property	Mars	Earth
Mean diameter (km)	6788	12,370
Mean distance from the sun (km $\times 10^6$)	227.8	149.5
Distance at perihelion (km $\times 10^6$)	206.5	147.0
Distance of aphelion (km $\times 10^6$)	249.1	152.0
Siderial period (tropical year)	1.88089	1.00004
Axial rotation period (solar day)	24 hr: 37 min 22 sec	24 hr
Axial inclination (degrees)	23° 59′	23 ° 27′
Surface gravitation	3.73 m/s^2	9.82 m/s^2
Mass (kg)	6.3×10^{23}	5.98×10^{24}
Mean density (g/cm^3)	3.933	5.52
Orbital period (s)	5.94×10^7	3.15×10^7
Spin rate (rad-s^{-1})	7.08×10^{-5}	7.27×10^{-5}
Semi major axis (m)	2.28×10^{11}	1.50×10^{11}
Eccentricity	0.0934	0.0167
Obliquity (deg)	25.1	23.5
Longitude of perihelion (deg)	70	102
Solar constant ($W \cdot m^{-2}$)	586	1360
Planetary albedo	0.23	0.30
Effective planetary temperature (K)	212	256
Planetary mean surface temperature (K)	218	288
Surface pressure (Pa)	700	1.01×10^5
Gas constant, R ($J \cdot kg^{-1} K^{-1}$)	189	287
Surface density ($kg \cdot m^{-3}$)	0.017	1.21
Atmospheric specific heat, C_p ($J \cdot kg^{-1} K^{-1}$)	830	1000
Adiabatic lapse rate, g/C_p ($K \cdot m^{-1}$)	4.5×10^{-3}	9.8×10^{-3}

Constituents	Mars	Earth
Carbon dioxide	96%	0.03%
Nitrogen	2.5%	78%
Oxygen	0.1%	21%
Argon 40	1.5%	0.9%
Argon 36	4 rpm	32 rpm
Water vapor	0-85 rpm	up to 5%

4

HISTORY OF THE WATER ON MARS

4.1 EVIDENCE OF WATER ON MARS

Many, many scientists examined the ten Shergotty-Nakhla-Chasigny (SNC) meteorites that are widely believed to be igneous rocks from Mars, ejected from that planet by one or more impact events. Without going into the many explanations that support the thesis that these samples are from Mars, let us accept for the moment the preponderance of experimental evidence that these rocks are Martian. It is assumed by most geoscientists that earth and Mars had very similar ratios of deuterium D, and hydrogen H. The water in the present Martian atmosphere is strongly enriched in D relative to water on earth, having a D/H ratio approximately five times the earthen value, which means that some hydrogen has disappeared from Mars. Examination of the amphiboles in the rocks shows they contain approximately one-tenth as much water as expected. That suggests that SNC magmas were less hydrous than was previously thought. Since the SNCs are medium-grained igneous rock that show evidence of crystal accumulation, they were probably shallow intrusions into the crust of Mars or perhaps surface flows of many crystals. Evidence from examining the clay minerals shows strong evidence that three of the ten SNCs interacted with fluids in the Martian crust at low temperature after the magma solidified. The samples show water that is "significantly enriched in D," revealing that the Martian crystal water had become D-enriched through its proximity to and interaction with the atmosphere.

Some of the crystals (notably kaersutite) contain hydroxyl (OH) as one of their major parts. Some contain 1.5% water (by weight).

Models of the evolution of the Martian atmosphere should include a large and exchangeable crystal reservoir of water. To change the D/H value of a large reservoir of water, a very large amount of hydrogen must have escaped from Mars. For Mars to have the gigantic amounts of water that has been suggested by some geophysicists, the escape rate of hydrogen must have been higher in the past, since 10^8 to 10^9 years ago, the D/H ratio was very nearly that of the present Martian atmosphere. Note, absorbed water has a low D/H value.

Figure 3 is a NASA-Johnson photograph of a type of meteor found in Antarctica tagged as ALH84001. Initially it was a type associated with the asteroid belt between Jupiter and Mars. Upon close examination in a laboratory, it was reclassified as an SNC, a type of meteorite containing rare gases that can be found in the Martian atmosphere. It was found to have come from Mars, probably ejected into space from an asteroid impacting on Mars approximately 15 million years ago. Using carbon-14 dating techniques, it had landed in Antarctica only 13,000 years ago.

Analysis revealed the rock was about 4.5 billion years old, about the age of earth. The rock contains carbonate globules (the nearly circular parts of the rock in Figure 4), iron oxide crystals (magnetite), and two types of iron sulfide. These are minerals that have been found on earth by bacteria.

Further analysis of the rock by laser mass spectrometry revealed an <u>organic</u> molecule called polycyclic aromatic hydrocarbons (PAHs). It is known that PAHs are easily created from organic molecular combustion (toasted marshmallows, for example). PAHs are also the trademark of flora and fauna in fossilized rock.

Some scientists suggested that the PAHs could be the product of bacteria. If so, then that finding would mean that life once existed on Mars.

Further scientific inquiry into the rock revealed tiny rounded objects that might be the fossils of minute Martian organisms (see Figure 5). These objects, however, are about half the size of the smallest known bacteria on earth.

These are exciting discoveries, yet are inconclusive as to the question if life ever existed on Mars. Skeptics, and there are many who are experts in their field, feel that though the findings are of enormous curiosity, the findings do not constitute proof: just strong likelihood. Many scientists, however, are on record that they would be very surprised if we later proved there never was life on Mars.

Figure 3. Photograph of Martian meteor ALH84001. Scale is in centimeters. Cube E is 1 cm^3. Photo courtesy of NASA Johnson Space Center, Houston, TX.

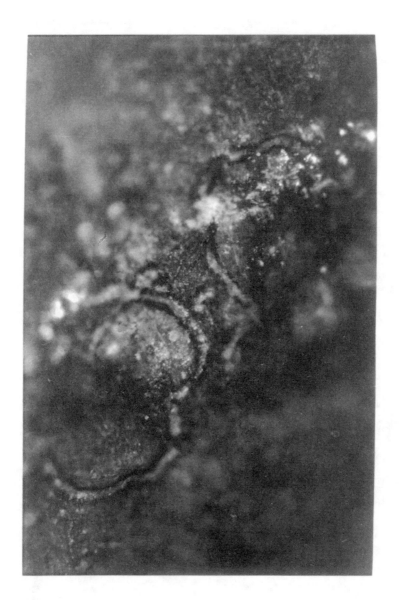

Figure 4. Electron microscope image of ALH84001 taken by Dr. Monica M. Grady, The Natural History Museum, Department of Meteorology, London. Courtesy of NASA, Houston, TX.

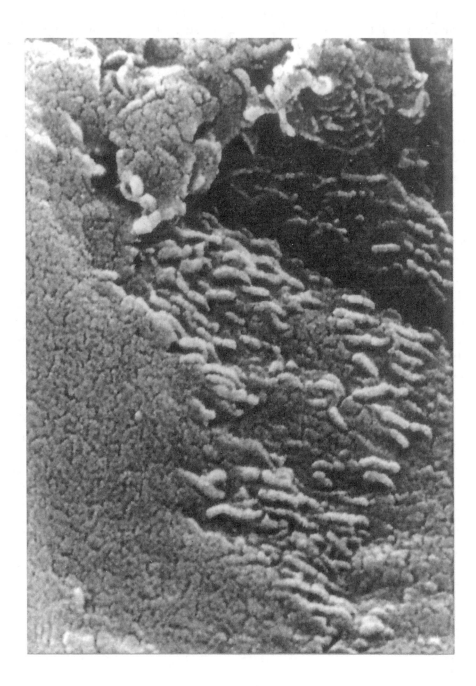

Figure 5. Electron microscopic image near center of specimen in Figure 3. The tube-like structures may be microscopic fossils of primitive bacteria-like organisms from Mars about 3.6 billion years ago. Fossils are 1/100 the diameter of a human hair. Photo courtesy of NASA Johnson Space Center, Houston, TX.

The chemistry of the carbonate material found in ALH84001 seemed to have formed at temperatures that could not sustain life. Some contend that the PAHs are the result of mineralogical processes; that the crystals were possibly formed by volcanic gases far too hot to support life. And then there are those scientists who believe that any presence of organic materials is due to contamination from earth when it arrived 13,000 years ago.

On the other side of the ledger are scientists who state that the carbon isotopic composition in the rock originates from organic production of methane gas. Examination of other tagged rocks from Mars suggests that life could have existed on Mars as recently as 500,000 years ago, which suggests that life could presently exist on Mars in certain areas.

4.2. THE MARTIAN CHANNELS

One of the major treatises on the history of water on Mars is Squyres [2]. The material in this section is taken from References 1 and 2. Squyres considers the water history as both a long term evolution and a short term. Photographs of Martian channels presented strong evidence that some of the channels were clearly the result of fluid flow, notably liquid water. The channels possess different morphologies, but those that possess a fluvial origin are grouped as either <u>outflow</u> or <u>runoff</u> channels[9]. The basic difference between the two is that the outflow channel is somewhat confined and its geometry is definitely bounded (see Figure 6). Here the outflow channel is represented by a sinuous multichannel source containing discontinuous marginal terraces and teardrop-shaped islands (blunt ends face upstream). The character of this particular channel indicates it might have evolved from fluids, such as from melting ground ice or permafrost. Runoff channels have a marked tributary system somewhat similar to our terrestrial river systems and are typically much smaller than shown in Figure 7. These channels were formed where the outflow channel flow was restricted and incised through ancient cratered terrain. A third type of channel exists on Mars, called fretted channels, and is related to ground ice (see Figure 8). We shall discuss each of these three channels and other phenomenon associated with the morphology on Mars in order to grasp some of the viewpoints on the history and distribution of Martian water. Let us consider each of these channels separately.

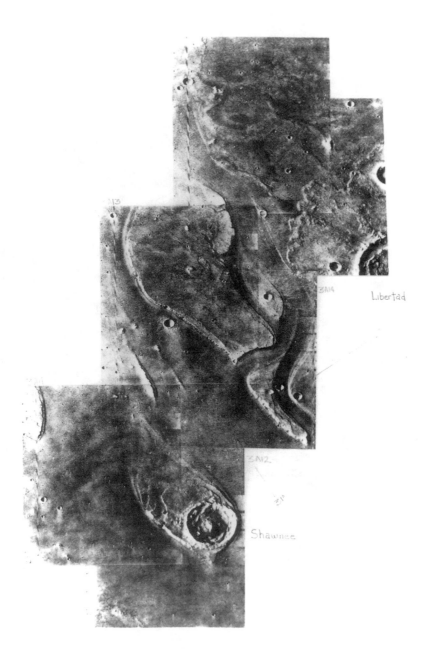

Figure 6. Spacecraft photograph of Martian outflow channel containing terraces and teardrop-shaped islands.
Courtesy NASA.

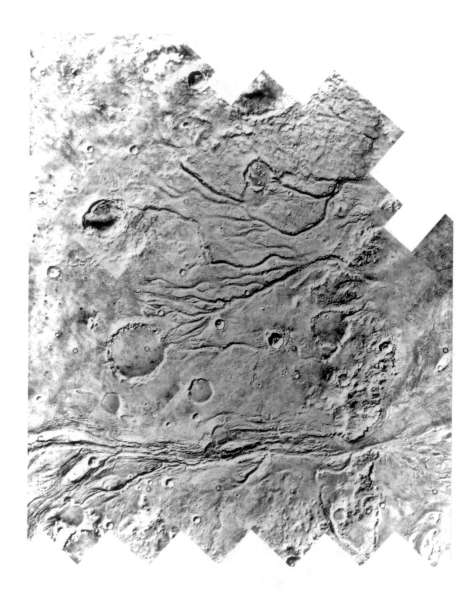

Figure 7. Spacecraft photograph of Martian runoff channels. The scale across the image is 265 km. Courtesy NASA.

Figure 8. Spacecraft photograph of a Martian region of fretted terrain. The terrain appears like a mosaic of mesas called lobate debris at the base of each escarptment. Courtesy NASA Mariner 9.

4.2.1. Outflow channels

Figure 6 is a typical outflow channel that arose from a chaotic terrain. It is similar to the type of formation caused by catastrophic flooding on earth. Features of Martian outflow channels and earthen flooded terrain that are common are:

i)   anastomosis
ii)   indistinct fluid sink features
iii) erosion of rock types hundreds of kilometers from their source
iv)   residual uplands separating the channels and streamlined by the flow
v)   flow constrictions and expansions
vi)   bar complexes below expansion points
vii) high width/depth ratios
viii) low sinuosity
ix)   differential erosion controlled by lithology
x)   longitudinal grooves
xi)   inner channels
xii) cataracts
xiii) pendant-shaped streamlined obstacles
xiv) scouring around obstacles

In all these cases, the Martian features are several times larger than the terrestrial equivalent. All the evidence points out that the Martian outflow channels were formed by catastrophic release of liquid water from confined aquifers. Orders of magnitude reach 5×10^8 m³/s (Carr [3]). Examination of the scouring concluded the principal erosion mechanisms were large scale turbulence, streamlining and cavitation (Baker [4]).

We should ask what initiated such large flooding, and several investigators recommended a few mechanisms. McCauley, et al. [5] and Masursky, et al. [6] believed volcanic eruptions beneath glaciers could produce large magnitude flooding (though no evidence appears for volcanic activity at the source of the outflow channel). Milton [7] recommended pressure release dissociation of large amounts of subsurface CO_2-H_2O clathrate, however a thermodynamic analysis does not support this viewpoint. Another possibility, proposed by McCauley et al. [5] and Sharp and Malin [8], suggested geothermal warming of subsurface ice in the chaotic terrain resulted in the flooding. If the geothermal warming is by means other than volcanic, this latter suggestion appears the most probable.

One of the major difficulties in explaining outflow channels is the accounting for the volume of material removed by the flood discharging. Carr [3] offered a very sound explanation. He suggested that the water was released at high pressure from an immense aquifer. He proposed that early in Mars' history its crust may have been brecciated and porous due to meteorite bombardment. This would permit large volumes of water to be stored beneath the surface. As Mars cooled, cold surface temperatures would cap the subsurface aquifer with an impenetrable permafrost. Then by thickening of the permafrost layer, warping of the surface layer could take place, creating very high pore pressures in the low lying warped areas. Then when the pore pressure exceeded the lithostatic pressure, hydraulic breakout would occur with huge discharges from the aquifer. Note, this explanation does not require any volcanic or geothermal activity, but it does require a very high permeability.

It is perhaps worthwhile to mention some other proposed mechanisms for chaotic flooding. Nummedal [9, 10] believes high water pore pressures forces subsurface liquefaction of clay-rich material releasing highly mobile debris flows, a similar mechanism to that proposed by Carr above. Lucchitta, et al. [11] suggested that enormous glaciers carved the outflow channels. Though this has great appeal, the origin of these glaciers is difficult to imagine since the glaciers of the type and size cited by Lucchitta are only found at the perimeter of immense ice sheets such as those found in Antarctica and Greenland.

In conclusion, the evidence thus far obtained appears to indicate that rapid flow of huge volumes of liquid water along with entrained debris such as rocks, sediment, and ice were responsible for the formation of the outflow channels.

Malin [12] and Masursky et al. [6] determined the age of the outflow channels to be younger than the cratered highland from which they arise. Malin used the impact flux model of Soderblom et al. [13] to estimate the outflow channels and found they are some 4 billion years old. Masursky et al. [6], using the same math model, estimate the age to be 2.5-1 billion years. Regardless what the model used, it can be stated that the outflow channels occurred fairly early in Martian history, after the formation of the southern highlands and crater bombardment.

4.2.2. Runoff channels

A second type of channel found on Mars due to the flow of water is the runoff channel. A typical system is shown in Figure 7. In one sense, they are not really channel-like open channel flow of a moving fluid. There is no evidence of streamlined obstacles, bars, interior channels or meandering properties typically found in terrestrial drainage systems. Thus, though they are popularly identified as runoff channels, they should be correctly called valley systems.

The runoff channels differ from terrestrial channels in a number of ways (Pieri [14]). The Martian channels have steep cliff-like walls, talus slopes, flat floors, and rounded basins to tributaries. They do not show dendritic patterns typical in terrestrial tributaries, but show instead parallelism. There is usually scanty runoff erosion on the surfaces, indicating negligible rainfall. Thus, rainfall was not the cause of the carving of most of these channels. The Martian channels are typical of channels formed by ground water sapping (see section 5.6) and liquid runoff. While the majority of runoff channels appear to have been formed by ground water sapping, precipitation cannot be entirely ruled out as a possible cause. All of the runoff channels[10] are found in the cratered highlands, formed more than 4 billion years ago, and thus older than the outflow channels, but still less than the bombardment period. Though their formation is almost certain to be due to liquid water, the actual discharge rates are much smaller than those associated with the outflow channels. Present conditions permit water to flow considerable distances on Mars providing it is protected from the surface temperature by a thick layer of ice (Wallace and Sagan [15]). Formation of the runoff channels by sapping implies the existence of subsurface liquid water at very shallow depths requiring high surface temperatures. (This will be discussed in depth in Chapter 5.) The runoff channels therefore present strong[11] evidence that the surface atmospheric temperature and pressure were considerably larger than they are today.

It is important to note that subsurface ice cannot presently exist in equilibrium with the atmosphere at latitudes lower than 40°. Any ground ice below these latitudes must, therefore, be isolated from the atmosphere by burial. A separate section will discuss subsurface ice in great detail. However, it should be noted that both the chaotic terrain and the fretted terrain give evidence to one manner of removal of subsurface ice. There were probably others. Figure 8 is a photograph of what a fretted terrain looks like, and Figure 9 depicts a chaotic terrain. Figure 9 shows a chaotic terrain with flat-floored steep-walled features: characteristics of a fretted terrain. Note the arcuate slump blocks at the lower edge of the chaotic area (arrow). The flat-floored chasm leading to the top may have been widened by the recession of walls or carved by a huge flood which burst from the area of chaotic terrain. Sharp [16] describes a fretted terrain as smooth, flat lowland area separated from a cratered upland by abrupt escarpments of a complex plan metric configuration. They appear as enormous mesas or buttes scattered randomly on a relatively flat bed. They are formed largely along the boundary of the northern lowland or southern highland regions. Sharp suggested that the escarpment recession resulted from evaporation of exposed ground ice or emergence of ground water. Regardless which, an enormous quantity of water was involved. Exactly what happened to the material debris that carved out these mesas is a mystery. It is possible that eolian activity transported it or it was uniformly distributed in an enormous settling basin.

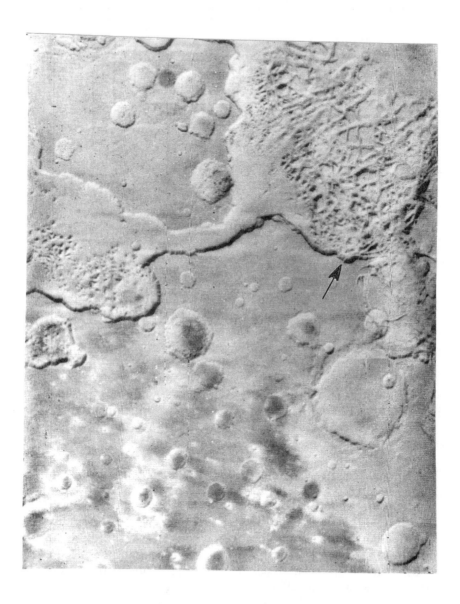

Figure 9. Mariner 9 photograph of chaotic terrain. Arrow points to arcurate slump blocks at lower edge of chaotic area. Courtesy NASA.

5

GROUND ICE

5.1 BACKGROUND

The geomorphic features of Mars suggest the existence of ground ice. By careful examination of escarpment heights we can gain some clue as to the physical mechanism causing the disappearance of ground ice. In studying the vast volumes of literature on the subject, there is continual reference that the investigator exercise extreme caution using terrestrial analogies to describe Martian features. However, there are features on Mars that have equivalent features on earth. For example, in examining depression and collapses of terrain, one has to discuss the phenomenon of thermokarst. Thermokarst is the process of melting ground ice accompanied by the collapse of the ground surface and the formation of depressions.

The principal features of the Martian fretted terrain appear similar to photographs of Alaskan Arctic thermokarst terrain. A possible explanation of the fretted terrain shown in Figure 8 is the collapse of upland cratered terrain due to the degradation of massive ground ice or subsurface ice. Figure 10 is an aerial photograph of a road passing through an area of ice wedge polygars. Terrestrial thermokarst topography consists of valleys, closed depressions, and basins with numerous frozen lakes. The perimeters of the polygars of Figure 10 are underlain by massive ground ice (ice wedges). Heavy dust (light-colored areas) and thermokarst (dark-colored parcels) are along the road margins. The thermokarst extends up to 20 m on both sides of the road. As the ice melts, the surface collapses.

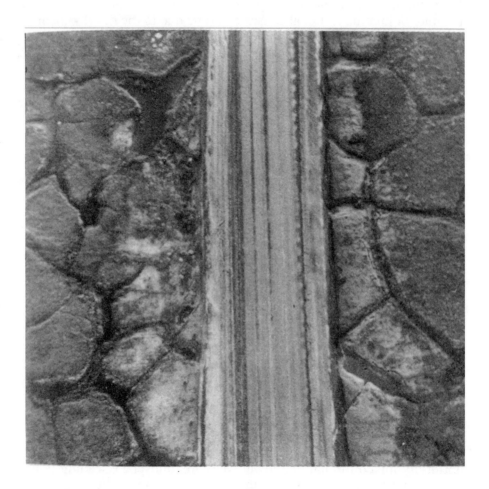

Figure 10. Aerial view of a road passing through an area of ice-wedge polygars in Alaska. Thermokarst are dark-colored patches. Reprinted with permission from *Science*, Vol. 238, No. 4828. Copyright 1987, American Association for the Advancement of Science.

Thermokarst occurs in places having a high ice content in the soil, which is one of the reasons it may occur in the lowlands on Mars. It is due to the disruption of the thermal equilibrium of the permafrost and increases in depth of the active layer. There are two reasons for disequilibrium and permafrost degradation: climatic and local. The climatic reason for permafrost degradation is, among others, due to an increase of temperature or humidity. Local causes could be due to the development of the polygonal ground. Accumulation of water occurs where ice veins cross or in low centered polygons (Hussey and Michelson [18]). The water temperature in these basins is always higher than the neighboring ground temperature. Due to thermal and chemical effects of the water, the ice veins start thawing and thermokarst begins.

On earth, the extent of the thermokarst process depends on (1) the tectonic regime of the area, (2) the soil's ice content, and (3) the degree and rate of the increase of the depth of the active layer.[12] The distribution of ground ice for thermokarst development can be of three types. The <u>first type</u> comprises ground ice in large deposits of stable regions. In these deposits, most ground ice occurs in the upper part due to water migration towards the freezing front. The <u>second type</u> comprises deposits where the ground ice is distributed throughout the entire vertical profile. The <u>third type</u> comprises ice cores of pingos.

There appears a large consensus that ground ice is the single most prospective reservoir for Martian water.[13] An estimate of its volume and composition would be a good indicator of Mars' water content. There are many investigators who believe there is ample evidence that Mars once had vast quantities of liquid water (see Squyres [1]). Due to cooling of the Martian climate, any water existing below the surface would have frozen, creating what we now call ground ice.

Based on Viking data, Farmer and Doms [19] made three assumptions in calculating zones in which ice could exist year round in thermodynamic equilibrium with the present Martian atmosphere: (1) that the water was uniformly mixed through the lower Martian atmosphere leading to a frost point temperature of 198 K, (2) the subsurface temperatures in the upper 10 m of the regolith was that given by the thermal mode of Kieffer et al. [20], and (3) the subsurface ice is in diffusive contact with the atmosphere when the atmospheric frost point temperature of 198 K was applied to ice stability throughout the regolith.

The result of Farmer and Doms' calculations is shown in Figure 11. The slashed region shows the latitude and depth where the temperature is less than 198 K, allowing ground ice to be stable all year. Note that for their assumptions the ice is stable at the surface from the poles uniformly to a depth of at least 10 m. Beyond ± 85°, the stability is somewhat linear to a depth of 1 m at about ± 60°, then drops off steeply to 10 m at ±

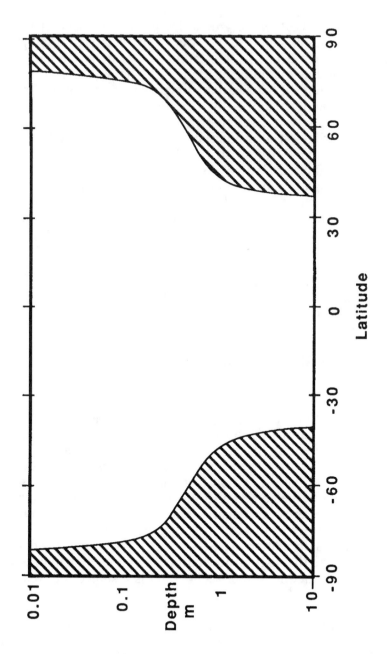

Figure 11. Martian zones in which ice can exist year round in thermodynamic equilibrium with the Martian atmosphere. From Farmer and Doms [19], copyright by The American Geophysical Union.

40°.[14] In the regions outside of the slashed region, temperatures exceed 198 K sometime during the year, resulting in ground ice being unstable and possibly lost into the atmosphere by sublimation and diffusion.

Figure 11 does not reveal where ground ice presently exists. The figure simply defines zones of stability. We know, for example, for temperatures greater than the atmospheric frost point, that ground ice can exist indefinitely providing there is no diffusion or sublimation. We also have no evidence of the history of the Martian climate so we cannot use any of these results with certainty. Obviously, variations in the planet's orbital and rotational dynamics would seriously affect the Martian climate and subsequently the distribution of ground ice. Others who studied ice loss are Fanale et al. [21] and Clifford and Hillel [22]. Fanale et al.'s model includes many physical effects, such as Knudsen and molecular diffusion, geothermal heat transfer, solar luminosity, climatic variations, and changes in albeido. One of their results is shown in Figure 12, where they calculate the long-term evolution of ground ice. The curves show the retreating position of the ice front with time. We see that the ice is gradually lost, which is due largely to the effects mentioned above.

Though we have a number of mathematical models that predict the evolution of ground ice, little is known regarding where the ice actually exists. The next section discusses this.

5.2 WHERE DOES GROUND ICE EXIST?

One way we can determine where ground ice may exist is to examine the morphology of Mars. Subsurface ice certainly can alter the geography of a planet's surface. It can melt and flow in streams, or flood, carve out meandering channels or enormous canals. It can move as a glacier cutting out huge channels. It can thaw, then freeze, resulting in polygonal ground patterns, as described in the section on thermokarst. It can be in a solid state then liquefy due to meteor bombardment. We can study channels, depressions, surface patterns, and distribution of rock debris to gain insight into the physics of water flow, be it in a liquid or solid state. Observation of solid state deformation on Solar System bodies has led to a rekindled interest in the rheology of ice, particularly low temperature ice. This subject will be treated in a later section and is a subject of great research interest.

Water ice is known to experience ductile deformation at very low temperatures. On earth, it has been observed to mobilize highly comminuted erosional debris that it cements. (See Durham et al. [23], Kirby et al. [24], and Kirby et al. [25]. They discuss the rheology of pure ice.) Knowing the thermodynamics of low temperature ice, one can examine Martian landforms formed by ice, due largely to viscous creep.

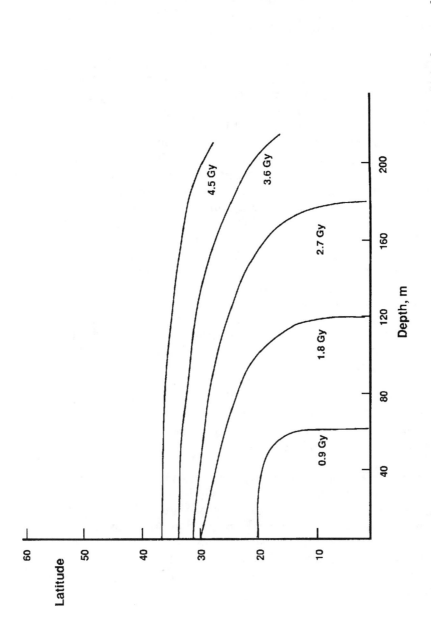

Figure 12. Long-term evolution of Martian ground ice. Curves give position of retreating ice front with time in Gy. Note, ice is gradually and irreversibly lost from near-surface equatorial regions. From F.P. Fanale et al. [21]. Permission granted.

Rock-ice masses, such as terrestrial rock glaciers, deform and flow due to solid state creep of the ice with no thawing taking place. The speed is considerable. Typical viscosities for terrestrial rock glaciers range from 2×10^{14} P to 9×10^{15} P (White [26]).

There appear to be three classes of landform that are the result of viscous creep: (1) that formed of solids which have been transported down an escarpment and lie at the base, typical of terrestrial gravity-induced landfalls, avalanches, and such. There is considerable debris accumulation at the bottom of escarpments on Mars with only a fraction appearing to be ice-cemented. Squyres calls these accumulations lobate debris apron. Figure 13 is a typical photograph taken by Viking Orbiter. The arrows indicate the lobate debris aprons. (2) The second class is termed terrain softening and is dissimilar to the deformation caused by thermo-karst or permafrost.[15] The lobate apron debris show morphologic evidence that a flow once existed. Figure 14 shows a small impact crater partially filled from a flow that topped a neighboring crater rim (arrow). On closer examination, one sees in Figure 15a, b that the surface was subjected to intense eolian erosion along with unstable ice. The Martian surface is filled with eolian landforms. (3) The third class of landforms that may give evidence of water flow is concentric crater fill, a phenomenon perhaps resulting from compression stresses due to the inward flow of ice-rich material from the crater walls. Figure 16 shows the distinctive concentric pattern of ridges and troughs as well as the likely eolian surface sketching of the deposits.

After careful study of high resolution photographs from the Viking orbiter, Squyres concludes that terrain softening is the result of viscous creep of near surface material, and for the lobate debris aprons, the cause of the creep is deformation of ground ice.[16]

As to the distribution of the above creep features, it can be stated quite reliably that no creep features are found within 30° of the equator. In the northern hemisphere, the lobate debris aprons are not widespread but localized in such regions as Tempe Fossae, Mareotis Fossae, the Phlegra Martes, and north of Olympus Mars near an old cratered terrain. Terrain softening, on the other hand, is found in all areas north of 50° latitude in the ancient cratered highlands. South of −30° latitude, there is quite a lot of terrain softening.

Calculations indicate that where ground ice is stable close to the Martian surface, then creep features are in great abundance. When ground ice is unstable, then there is an absence of creep features. Thus, if creep features result from ground ice, then their distribution gives good evidence that there is abundant ice near the surface at high and middle latitudes and little ice at low latitude.

Figure 13. Viking orbital image of lobate debris aprons (arrows) revealing the convex topographic profiles.
Courtesy of NSSDC, M.H. Carr, Team Leader.

Figure 14. Viking orbital image of morphologic evidence of flow. Arrow points out one flow that has over-flowed into a crater. Courtesy of NSSDC, M.H. Carr, Team Leader.

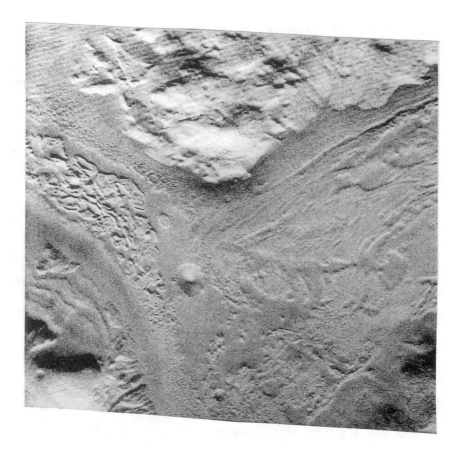

Figure 15a. Viking orbiter image of a lobate debris apron and lineated valley fill material. Note the deeply etched appearance of flows. Courtesy of NSSDC, M.H. Carr, Team Leader.

Figure 15b. Viking orbiter image of a lobate debris apron where etched appearances suggest surfaces had significant eolian erosion. Courtesy of NSSDC, M.H. Carr, Team Leader.

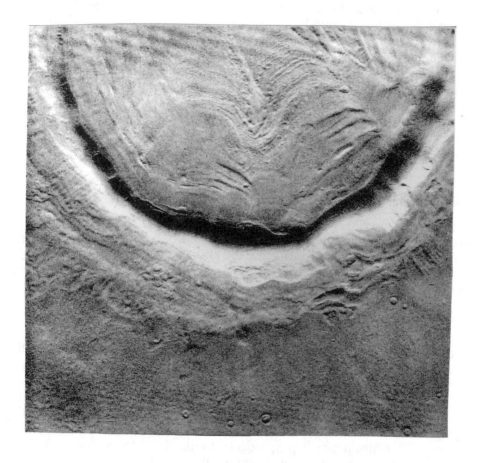

Figure 16. Viking orbiter photograph showing a crater fill. Note the small-scale eolian etching of lineation.
Courtesy of NSSDC, M.H. Carr, Team Leader.

Having established somewhat some ground rules for the existence of ice, it is worthwhile to consider at what depths ice may lie. One of the basic issues Squyres confronts is the depth to which the deformation extends in terrain softening. The mathematical model selected was the MANTLE finite element code of Thompson et al. [27, 28], who solve the Navier Stokes equations for incompressible creeping flow. The calculation was performed for a crater form. Symmetry is rotational about the crater's center. The boundaries are a free upper surface and a rigid lower surface. The boundary conditions are the no slip condition at the bottom surface and unconstrained velocities at the upper surface. The deforming layer has variable thickness, and the fluid is assumed Newtonian. Though the model cannot predict the depth of ice, it does reproduce the morphology of impact craters observed in the softened terrain on Mars. It also shows the thickness of the deforming layer to be a small fraction of the diameter of the crater. The conclusion is that terrain softening is due to viscous creep and is a near surface phenomenon.

5.3   THE RHEOLOGY OF MARTIAN GROUND ICE

In treating the rheological description of Martian ground ice, one must consider three factors that influence the strain rate of frozen ground experiencing steady-state creep: stress, temperature, and grain size of the material. We shall consider each separately.

5.3.1   Stress dependency

Stress dependency depends on diffusion and dislocation movement. Diffusion creep involves the diffusion of atoms along grain boundaries that takes place at low temperatures and within individual grains which occurs at high temperatures. The deformation is Newtonian, and dislocation creep results from the motion of various types of dislocations through the crystal lattice and results in a non-Newtonian deformation. For dislocation creep, the strain rate is proportional to the stress to the power 3 using typical Martian regolith temperatures, the shear stress necessary for Newtonian diffusion creep is 10^5-10^6 degrees/cm^2. For higher shear stress, non-Newtonian dislocation creep occurs. Considering a soil and ice density of 1.8 g/cm^3 and a topographic slope of 3 degrees, a shear stress of 10^5-10^6 degrees/cm^2 was calculated by Squyres [1] to exist at a depth of 30-300 m. Anything above this depth (using the same data), the deformation would be Newtonian, and any greater depth, the deformation would be non-Newtonian. Thus, topographic relaxation could be dominated by non-Newtonian flow in the first stages followed by Newtonian flow. For non-Newtonian flows, we have less viscosity and greater strain rates at deep depths.

5.3.2 Temperature dependency

Temperature dependency of ice is fairly well complete. Only in the low pressures and −10° C to −40° C range are there gray areas where doubts exist. The rate of steady state creep is proportional to $\exp(-Q/kT)$ where Q is the activation thermal energy of the deformation mechanism, k is Boltzmann's constant, and T is absolute temperature. The temperature dependence of viscosity should play a significant role on the rheology of the regolith as a function of latitude. The exponential dependence of rheology on temperature suggests that creep is almost negligible at high latitudes. As pointed out by Squyres, "the present annual average surface temperature on Mars at 45° latitude where terrain softening is common is about 205 K. The cyclic changes in Martian climate, of which the most important are caused by obliquity variations, have essentially no effect on mean annual temperature at this latitude, so this temperature applies throughout the obliquity cycle. At latitude 75°, where terrain softening is uncommon and minor, a typical mean annual surface temperature at the present obliquity is about 160 K. At this latitude, however, the thermal effects of obliquity variations are substantial. At the lowest obliquity, the mean annual temperature drops to about 145 K, while at its highest obliquity it rises to 180 K."

The approximate viscosity of ice at the mid-latitude temperature of 205 K will be lower than the high latitude viscosity by factors of 10^2 (for 180 K), 10^5 (for 160 K) and 10^6 (for 145 K). It is easy to see, then, why terrain softening is so much less pronounced at high latitude: the material there is very much stiffer than at the lower latitudes.

Fanale et al. [21] give a good mathematical model that quantitatively accounts for (1) obliquity and eccentricity variations; (2) long term changes in the solar luminosity; (3) variations in the argument of perihelion (in planetocentric equatorial coordinates); (4) albedo changes at higher latitudes due to seasonal phase changes of CO_2; (5) planetary heating and heat flow; (6) temperature variations in the regolith as a function of depth, time, and latitude due to (1), (2), (4) and (5); (7) atmospheric pressure variations over a 40^4-year time scale; (8) effect of (1), (2), (3), and (4) on annual polar cap temperature which determine annual average atmospheric H_2O concentration; and (9) Knudsen and molecular diffusion of H_2O throughout the regolith. It is the single most powerful model found thus far.

We next turn to a discussion of the thermodynamic properties of ice. Figure 17 will be our reference, and though not all the regions shown in the figure exist on Mars, we shall discuss the assorted transition regions for completeness. It should be noted that regions II, III, and V do not exist on Mars.

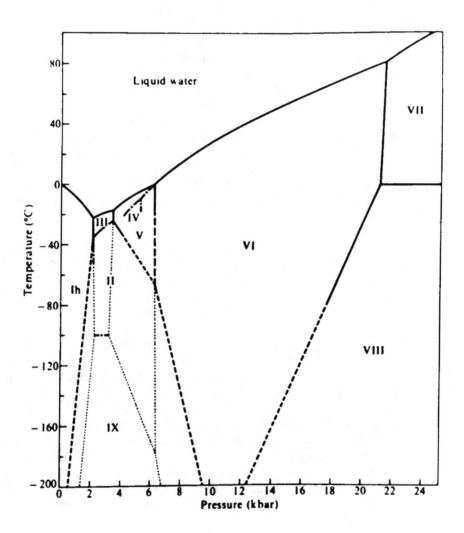

Figure 17. Phase diagram of ice.

The rheological equation for creep of ice V at temperature $-10°$ C to $-40°$ C, that gray area we spoke of earlier, is given by

$$\dot{\varepsilon} = 5.4 \times 10^{-2} \sigma^{2.7} \exp\left[\frac{-32400 + 10.1\,p}{RT}\right] \qquad (2)$$

where

$\dot{\varepsilon}$ = creep rate (s^{-1})
σ = shear stress
p = hydrostatic pressure (Mpa)
T = absolute temperature (K)

Table II is a collection of temperature, pressure, volume changes, entropy changes, and latent heats for the transition regions for ice. The data is taken from Bridgeman [29]. Values of Young's modulus Y, rigidity modulus G, Poisson's ratio γ and, Bulk modulus K of ice are found in [30]. Some typical values for polycrystalline ice at $-5°$ C are

$$Y \sim 9 \rightarrow 9.4 \times 10^{4} \text{ bar} \qquad (3)$$

$$G \sim 3.4 \rightarrow 3.8 \times 10^{4} \text{ bar} \qquad (4)$$

$$\gamma \sim 0.31 \rightarrow 0.365 \qquad (5)$$

$$K \sim 8.7 \rightarrow 11.3 \times 10^{4} \text{ bar} \qquad (6)$$

Reference 30 also contains stress versus rate of strain for a variety of ice conditions. In Table II, only transitions x, xi and xii are relevant to Mars.

An excellent reference for the physics and chemistry of ice is contained in the assorted papers of the VIIth Symposium on the Physics and Chemistry of Ice [31]. Some of the topics are i) the fractures of ice I_h; ii) inelastic properties of several high pressure crystalline phases of H_2O: Ices II, III, and V; iii) viscosity ice V; and iv) the role of the water layer at an ice surface in the kinetic processes of growth of ice crystals. Another excellent reference is [32], particularly a section dealing with the mechanical properties of ice II and III.

TABLE II. Thermodynamic Properties of Ice (from Bridgeman [29]).

	Transition	T(°C)	Pressure (k bar)	Volume change (10^{-6} m^3 mole^{-1})	ΔS (J mole^{-1} deg^{-1})	Latent heat (J mole^{-1})
i	$I_h \rightarrow II$	−35	2.13	−3.92	−3.15	−752.4
ii	$I_h \rightarrow III$	−22	2.08	−3.27	1.67	392.9
		−35	2.13	−3.53	0.67	167.2
iii	$II \rightarrow III$	−24	3.44	0.26	5.10	1270.7
		−35	2.13	0.39	3.85	919.6
iv	$II \rightarrow V$	−24	3.44	−0.72	4.85	1203.8
v	$III \rightarrow V$	−17	3.46	−0.98	−0.29	−71.1
		−24	3.44	−0.98	−0.25	−66.9
vi	$V \rightarrow VI$	0.16	6.26	−0.70	−0.04	−16.7
vii	$VI \rightarrow VII$	81.6	22	−1.05	~0	~0
viii	$VI \rightarrow VIII$	~5	~21	—	~ −4.22	−1178.7
ix	$VII \rightarrow VIII$	~5	~21	0 ± 0.0005	~ −3.89	−1086.8
x	$I_h \rightarrow liq$			−1.63	21.58	5,977
xi	$I_h \rightarrow vap$	0.01	6×10^{-6}	—	185.92	50,829
xii	$Liq \rightarrow vap$			—	164.34	44,851
xiii	$III \rightarrow liq$	−17	3.46	0.43	18.01	4,611
xiv	$V \rightarrow liq$	0.16	6.25	0.95	19.26	5,267
xv	$VI \rightarrow liq$	0.16	6.25	1.65	19.34	5,284
		81.16	21.5	0.59	17.93	6,354
xvi	$VII \rightarrow liq$	81.6	22.4	1.64	17.93	6,354

With regard to the subject of regelation, it is important to note that the change dT_m in the equilibrium melting temperatures $T_m(K)$ of solid ice due to a small change dp in hydrostatic pressure is given by

$$\frac{dT_m}{dp} = \frac{v_L - v_s}{S_L - S_s} \tag{7}$$

when v_L and S_s are volume and entropy of a unit mass of liquid and subscript s denotes the solid state, or

$$dT_m = \frac{T_m(v_L - v_s)}{L_f} dp = -A\, dp \tag{8}$$

where L_f is the latent heat of melting per unit mass, and $A = 0.00743$ °C bar^{-1} at 0 °C.

Figure 17 is a p-T plot of ice as presently known. The phase diagram is composed of a number of lines. The measured stable lines are denoted by ⎯⎯⎯⎯⎯⎯⎯ . The measured metastable lines are denoted by · ⎯⎯ · ⎯⎯ · ⎯⎯ · · . The extrapolated or estimated stable lines are denoted by - - - - - - - - - - . The extrapolated or estimated metastable lines are denoted by ·············· . Ice IV is metastable in the region of stability of ice V; the indicated field for ice IV is inferred from the D_2O system. Ice 1c and vitreous ice are not indicated in the figure. The triple point among I_h, liquid water, and water vapor is not shown since it occurs at too low a pressure to be shown on the same scale. We also see in the figure that the transition from III to II cannot be made. This is largely due to instabilities. Ice V is the most difficult to create in the laboratory. Ice I_h displays a rich variety of inelastic behavior at low to intermediate temperature, behavior that stems from a range of deformation processes that are poorly understood. We need more exploratory experimentation to better understand the rheology of this intriguing material at low temperatures. Again, ices other than i are not relevant to Mars.

Ice I_h displays a frictional law [33] of the form

$$\tau_f = 8.5\,\text{MPa} + 0.2\sigma_n \ (\text{MPa}) \tag{9}$$

where τ_f is the shear stress necessary to overcome static friction and σ_n is the normal stress at the sliding surface at the time that sliding occurs. Friction at these conditions is unstable, producing rapid stress drops that can take place repeatedly upon reloading providing no rupture occurs. The above equation is insensitive to temperature over the range 77 K $\leq T$ $\leq$ 115 K and sliding velocities range from 3×10^{-4} mm/s $\leq \bar{V} \leq 3 \times 10^{-2}$

mm/s and 20 MPA $\leq \sigma_n \leq$ 250 MPa. The micromechanical mechanisms controlling ice friction are completely unknown at this time.

5.3.3 Grains

Grain size (structure) is the third and perhaps most important factor in the rheology of an ice-material composition, i.e. the structure of the mix or the proportion of ice to material and how one interacts with the other. Obviously, the composition is crucial when considering the kinematics of the mixture, as the grain size can be negligibly small so the mixture behaves as pure ice and is highly mobile, or it can increase to the size of rocks and be relatively immobile. Grain size is thus considered primary in the determination of ice depth on Mars.

For <u>small grains,</u> or large grain at low concentration, we consider the particles as suspended in ice and increasing the mixture's viscosity. Theoretical models exist that describe the viscosity of such mixtures (Thomas [34]). "Once the particle volume concentration exceeds 0.5-0.6, simple theory breaks down because the particles begin to lock together." Numerous experiments have been performed with frozen soils where ice creep occurs at concentrations as high as 0.8 (see [35], for example). In such high concentrations, the ice acts as a lubricant, resulting in less frictional grain-to-grain resistance to flow.

In conclusion, all these factors (stress, temperature, and structure) affect the rheology of the Martian regolith. Stress and temperature increase mobility with depth at a given latitude. Structural effects, on the other hand, decrease mobility with depth, causing the flow to be stationary at depth where the major factor is solid rock, not ice.

5.4 DISTRIBUTION OF ICE AS DEDUCED FROM BOTH TERRAIN SOFTENING AND RAMPART CRATERS

Figure 18 is a photograph of a rampart crater. Viking 1 photographed the crater Yuty, located near the spacecraft's potential landing site, from a range of 1877 kilometers (1165 miles) on June 22, 1976. Yuty, 18 kilometers (11 miles) in diameter, has a central peak, and probably was made by the collision of a meteorite with the surface of Mars. The lobate flows are layers of broken rocks thrown out of the crater by the shock following impact. The leading edge of the debris flows forms a ridge similar to great avalanches on earth. The whole area has been worn down by wind and possibly water erosion that accentuates the surface detail. The rim of Wabash Crater, about 40 kilometers (25 miles) across, lies at the right

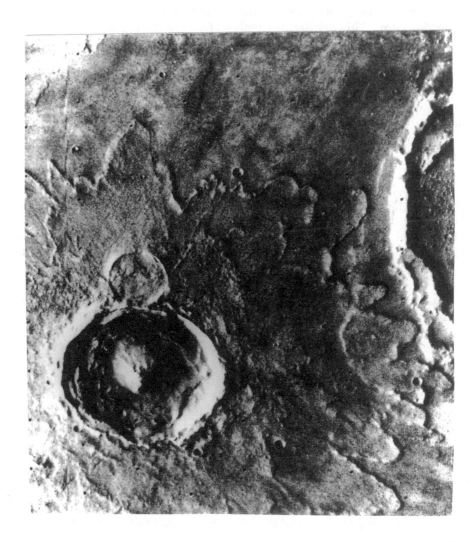

Figure 18. Viking orbiter photograph of the Martian crater Yuty. Courtesy of NSSDC, M.H. Carr, Team
Leader.

edge of the picture. Yuty was named for a village in Honduras and Wabash for a town in Indiana.

The rampart craters of the type shown in Figure 18 are a good indicator of subsurface water on Mars. Though the craters exist at all latitudes, water does not. This conflict is resolved by considering all the factors that play a role in making the Martian regolith's rheology.

Rampart craters are surrounded by lobed ejecta that have the appearance of a mud slide. The most common accepted viewpoint is that the impact melted and subsequently caused the subsurface ground ice to flow [36]. So if rampart craters and terrain softening are both indicators of ground ice, and we have stipulated earlier that terrain softening indicates ground ice at only latitudes greater than 30°, we have a conflict.

First, not all rampart craters have flow ejecta. As pointed out by Squyres [1], "there is usually some critical diameter below which ramparts are not observed and above which they are common." The obvious interpretation of the observation is that H_2O is present at depth, but absent close to the surface. An impactor must penetrate through the dry surface layer to the H_2O rich layer underneath for ramparts to form.

Kuzmin et al. [37] studied 10,500 craters over Mars noting the onset diameter for crater ramparts as a function of geographic position. Their results showed a striking latitude variation in the depth of water. For latitudes less than ± 30°, the depth to the H_2O rich layer was found to be from 300 to more than 400 m. At about ± 30°, the depth was found to rise abruptly to 200 m, and at ± 50° it rose to 50-100 m. The trend is thus clear. From studying the rampart craters, the result of subsurface ice indicates ice is close to the surface at high latitudes and is substantially deeper at low latitudes with a sharp drop at about ± 30°. This is schematically shown in Figure 19. The dashed line in Figure 19 indicates a transition from pulverized debris near the surface to fractured bedrock below the dashed line. The ordinate is not labeled since we do not know to what depth terrain softening extends or the depths of the ground ice with any accuracy. That will have to be accomplished by future spacecraft missions, establishing surface stations that can measure those parameters with great accuracy.

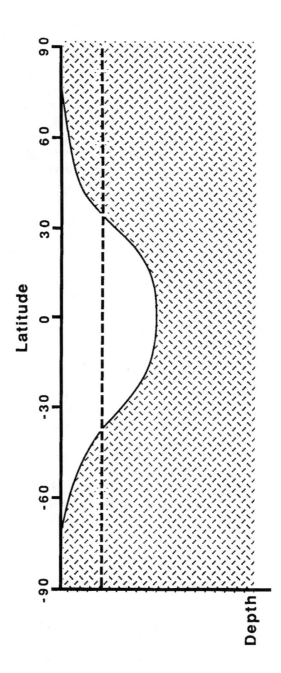

Figure 19. Pole-to-pole distribution of ice as deduced from terrain softening and rampart craters. (From Squyres [1]. Courtesy of ICARUS.)

5.5 GROUND WATER SAPPING

It is recognized that terrestrial stream valleys are formed by a combination of surface water and ground water processes. Kochel [38] investigated ground water sapping processes in laboratory experiments showing that they play a dominant role in the origin and evolution of valleys in physiographic settings that were similar to Viking orbital imagery of Mars. Though there appears little controversy in the importance of ground water sapping in the morphology of large valleys, there are a few scientists who believe that the climatic conditions on Mars were never similar at any time to conditions on earth. There are also significant problems of scale and time in the similarity between actual conditions on earth and laboratory experiments. The term ground water sapping is rather broad, referring to the weathering and erosion of sediments and rocks by emerging ground water in some combination of intergranular flow and channel flow. A typical discharge is a spring, if it is local, and seepage erosion if laterally extensive. Because it is difficult, if not impossible to relate terrestrial phenomenon to Martian phenomenon due to different climates, temperatures, material structure, and a total lack of any history of Mars, ground water sapping appears to be a remote phenomenon that created the morphological features of Mars. If, on the other hand, evidence existed that the climates were similar, then ground water sapping would be, in this writer's opinion, a viable mechanism.

What evidence does exist that ground water and ice existed until fairly recently (or still exist, as some suspect they do in the equatorial area) comes from evaluation of landslides, rock, and volcanic vents. Some claim the landslides were lubricated by water, as was mentioned earlier in the discussion of the lobate debris aprons. Luccitta [39] claims the material in one of the Martian channels was capable of erosion at considerable distances from its source, that it breached a bed rock ridge, carved flutes in the lower channel, and eroded its banks. There is strong evidence[17] that these landslide deposits contained fluids, that apparently water discharged from the canyon walls, if canyon tributaries were indeed formed by sapping.

5.6 CHEMISTRY OF ICE-COVERED LAKES

We might pose the question: What were conditions like in those temporary large basins of water early in Martian history? To answer that question, we have to know what the climate was. It would be wonderful to speculate that it was warm, warm enough to have a liquid state, even though it was exposed to a very cold atmosphere. We could imagine geotherms, underground water currents that kept a large part of the basin liquid even for very low atmospheric temperatures. We know that there can be significant erosion with a flow protected by an immense ice cover. There are a number of lakes in Antarctica that have liquid water under an ice sheet 3.5 m thick, and the Antarctic ocean exists for certain seasons under ice at extreme depths. Much of the sea ice is seasonal, and hence fairly thin (a few meters). The ice shelves are very thick since they originate glacially.

Ice cover has a pronounced effect on the chemistry, sedimentation, and biological activity of water (see [40], for example). The critical factor allowing lakes to remain liquid despite the atmospheric temperatures is the latent heat of the circulating water. Latent heat of water was discussed in Chapter 2. From Table II, a typical value of the latent heat of fusion is approximately 8×10^4 cal/kg, making water one of the best media for storing heat. If we now assume the ablation from the surface equals the freezing at the ice/water interface, then the ice cover will maintain an equilibrium thickness. That is, the thickness is constant because the conductive energy flux upward through the ice sheet exactly equals the heat removed by ablation at the ice surface plus the radiant energy absorbed by the ice. Or, stated in another way, any standing bodies of water are expected to have a perennial ice cover, such as those found in the Dry Valleys of the Antarctica. Here, the mean annual temperature is well below freezing, and ice has reached an equilibrium thickness. As ice is lost from the upper surface by ablation, new ice forms at the lower surface, releasing latent heat as it does. This heat is the dominant term in the energy balance equation that gives the equilibrium ice thickness. The thickness of the ice is presented in section 5.7.1.

Thick layered deposits are evident depressed areas (see [43, 44, 45]). They show fine, nearly horizontal layering and are viewed as isolated plateaus of what may have been extensive depositing. One investigator [46] believes the morphology of the deposits is due to standing bodies of water. Stating that ample evidence exists that the Martian regolith is highly porous and permeable and that it contained vast amounts of water at one time, Nedell considers four processes that could transport sediment downward through an ice cover: (1) solar energy warming individual particles, allowing them to melt through the ice, (2) sediment working its way downward through vertical melt channels, (3) layers of sediment deposited on the ice being thick enough to cause the ice layer to flounder, dumping the sediment into the lake, and (4) layers of sediment deposited on the ice cover leading to a Rayleigh-Taylor instability, and sediment diapirs penetrating downward through the ice layer similar to regelation. The various energy transfers are shown in Figure 20. The above energy balance gives a good estimate of ice thickness using measured values of ablation rate, 30 cm year^{-1} solar flux, ice albedo, and ice extension length. McKay [41] calculated values very close to measured terrestrial lake ice.

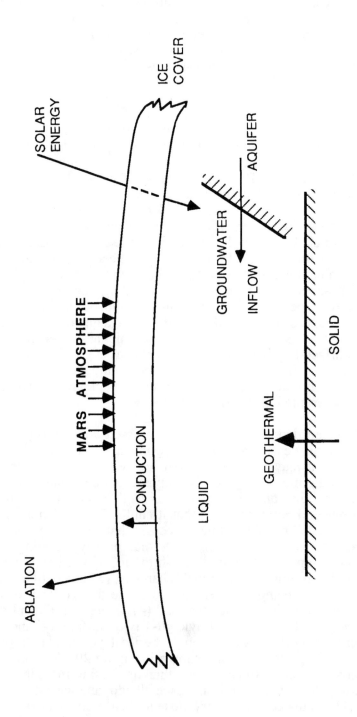

Figure 20. Energy transfers that take place on Mars.

Certainly, the ablation rate exerts a major effect on the thickness of ice, and for Mars, Squyres used Brutsaert's model of evaporation into planetary atmospheres [42].

5.6.1 Transport of Water by Ablation

Ablation from the ice surface is by sublimation to the Martian atmosphere at a rate which is controlled by the ice temperature and by the upward diffusion of water vapor. At the ice surface is a laminar sublayer that extends up to a fully turbulent layer. In this sublayer, vapor is transported from the ice surface as well as gases from eddies embedded in the turbulent layer. The thickness[18] of the sublayer is assumed approximately 1 m and will tend to be of the magnitude of the rough element heights. The water vapor being transported is controlled by molecular diffusion into Komologorov-scale eddies and then up into elevations by eddy diffusion. The transport is controlled largely by the stability of the atmosphere. Squyres considered three different atmospheric conditions for a neutrally stable atmosphere: (1) the present pressure of 7 mbar, (2) an early atmosphere of 300 mbar, and (3) an earlier dense atmosphere of 1000 mbar. A 300 mbar CO_2 atmosphere creates about 15° warming above the present temperature, and a 1000 mbar CO_2 atmosphere creates about a 30° warming, using the present mean annual surface temperature of ice as $T_S = 215$ K and the atmospheric temperature and ice surface temperature when ablation occurs as $T_a = 230$ K. At 300 mbar, the temperature becomes $T_S = 230$ K and $T_a = 245$ K, and for 1000 mbar, $T_S = 245$ K and $T_a = 260$ K. In all three cases, ablation is assumed to be confined to a midday period of 6 hr/day. Atmospheric water vapor content at the extreme boundary layer thickness is assumed to be 50% of saturation during the daily ablation period. The above analysis is very crude and valid only for a neutrally stable atmosphere. It is presented to show what is involved.

Squyres [1] calculated the ice ablation rates as a function of wind velocity at the top of the atmospheric boundary layer on Mars for the three sets of atmospheric conditions. Figure 21 shows the results of the calculation. For the present conditions on Mars, ablation rates are 1-2 cm per year, increasing to approximately 5 cm for the 300 mbar atmosphere and 10-20 cm for the 1000 mbar atmosphere. The corresponding ice thicknesses based on these ablation rates are shown in Figure 22. As can be seen from the figure, the greater the atmosphere the less the ice thickness. Using present condition data, the ice is hundreds of meters thick, whereas in ancient times, it could be only slightly more than a meter thick for all wind velocities, conditions found today in the Antarctic (though the solar input at the Martian equator is greater than it is in Antarctica).

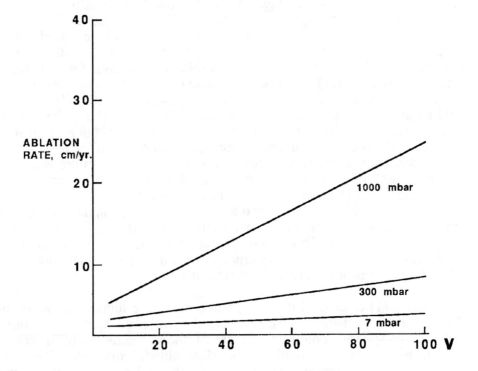

Figure 21. Wind speed at the top of Mars' atmospheric boundary layer. Speed is m/s. (From Squyres [1]. Courtesy of ICARUS.)

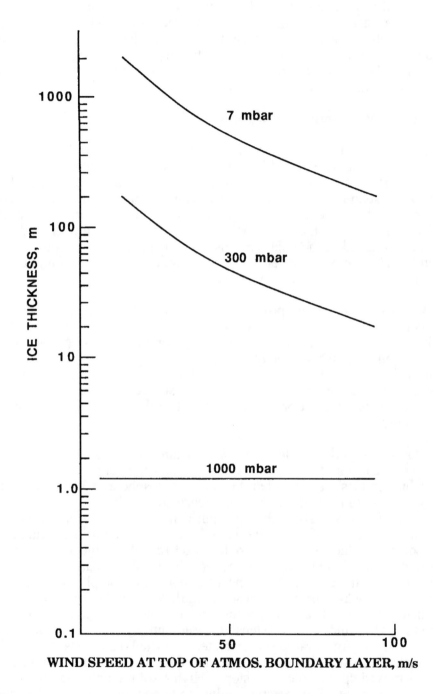

Figure 22. Equilibrium thickness of an ice-covered Martian lake for three atmospheric pressures. (From Squyres [1]. Courtesy of ICARUS.)

Under present conditions, a Martian lake has ice reaching to extremely deep depths. One cannot nor should not expect to find any liquid water at surface or near surface conditions. However, for a CO_2 atmosphere and with greenhouse warming in excess of 273 K, a lake can exist in equilibrium under an ice cover of tens of meters thick, providing the water can replenish the water loss by ablation. Again, that inflow must come from some subsurface aquifer.

5.6.2 Atmospheric Conditions

The composition of the present Martian atmosphere is difficult to interpret. For some scientists [47-49] Mars underwent outgassing that produced a layer of water 10-100 m deep over the surface of the planet. Besides ground ice, water could be stored in the present atmosphere, some could be lost to space by dissociation, some could be incorporated chemically in silicates, some could be adsorbed on regolith grains, and some could be trapped in the polar regions. The one common accepted conclusion is that there is no evidence that there are open streams of liquid water or atmospheric precipitation on Mars.

When we address the problem of water on Mars, we are actually addressing the issue where has the water gone, because too much evidence exists that water once was evident in huge amounts carving many of the features on Mars we see today. We have already stated much of it is frozen as ground ice. This we showed was based on morphological evidence. Indeed, ground ice constitutes the largest source of water on Mars. Of course, it is less than it once was, and part of that loss may be trapped in the atmosphere.

Mariner 9 [50] revealed the Martian atmosphere as having features similar to earth's atmosphere. Pressures and temperatures in the lower Martian altitudes are equivalent to earth's atmosphere at 30-40 km. Here, CO_2 is the dominant gas, and water freezes out. Though the relative humidity on Mars is fairly high, the vapor concentration is small in volume. Water-ice clouds do form as shown in Figure 23. Clouds appearing along the southern edge of the north polar hood reveal a large-scale turbulent wind pattern. Intense cold air covers the northern part of the region shown. Note the parallel bands near the bottom of the photo. This structure suggests waves produced in shearing flow along the bands, a structure formed in terrestrial satellite photographs of cold fronts and associated jet streams. One finds rather extensive cloud activity near the larger volcanoes on Mars. An infrared spectrometer identified the clouds as water-ice, somewhat at low altitude possibly caused by convection cooling as the clouds moved up the volcano's slope. High resolution photography shows cells of water-ice. The extreme cooling to temperatures of –127° C cause CO_2 clouds to form. Figure 23 shows large-scale wind patterns along the southern edge of the north pole hood. There appears a haze of CO_2 ice

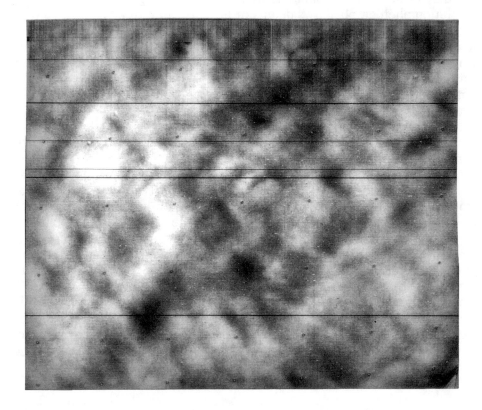

Figure 23. Mariner 9 photograph of clouds appearing along southern edge of the north polar hood revealing a large side turbulent wind pattern. (48° N, 6° W). Courtesy of NASA

crystals in the polar air close to the cold surface. This haze disappears in late winter to reveal a CO_2 frost or snow on the ground. Then between 45° and 55° latitude water-ice clouds form at heights up to 20 km. The wave pattern shown in Figure 23 indicates the wind direction is from the west at all elevations, with wind velocities from 10 m/s near the ground to 60 m/s at 10 km. The cloud patterns are similar to terrestrial photographs of cold fronts and jet stream patterns.

Another fascinating photograph is shown in Figure 24. This high resolution photograph shows details of the formation of a wave cloud over a crater in the north polar region. The wind is blowing from upper left to lower right, and a small wave cloud is forming about 40 km downstream. Both wave clouds appear to be quite turbulent. The diffuse appearance is caused by a wide thin haze of condensed CO_2 or H_2O. The layer crater is notably distinct due to either surface ice or snow (CO_2 or H_2O) around its rim.

The water vapor content in the Martian atmosphere was known to be very low. This was before Mariner 9 measurements. Some of the water probably has been photochemically dissociated to hydrogen and oxygen. So we ask: What happened to these elements? Hydrogen could escape from Mars' gravitational field due to its low atomic weight. Mariner 9 measured this escape rate. It was determined [51] that 2×10^8 atoms escaped per second every 1 cm^2 through an imaginary spherical surface at the top of the atmosphere. Assuming the only source of hydrogen was from water on Mars, this escape rate is equivalent to 1 million gallons of water lost from Mars each day. If this escape rate is assumed constant[19] since its formation, the total water that has escaped from Mars is roughly equivalent to covering the entire planet with water to a depth of 5 m. Certainly this is not a great depth, nor does it imply little to no water remains on Mars. Certainly water ice exists in the polar caps and as ground ice. Also, the hydrogen atoms could have come from the thermalization of solar wind protons (hydrogen ions).

The oxygen atoms need not have escaped from the planet. Some of the oxygen could have recombined with major ions in Mars' ionosphere, and considerable oxygen could have been used in surface oxidation.

The Viking lander data revealed that the water vapor is mixed evenly vertically throughout the Martian atmosphere and is not concentrated near the surface as on earth. Farmer and Doms [19] reported the distribution of water vapor over the planet for one year from the lander's atmospheric water detectors (MAWD). Figure 25 shows the distribution of water vapor in the atmosphere by latitude and season. The central dashed line shows the latitude when the sun is at its zenith at noon. The lined shaded areas represent the extent of the annual polar caps. Note the large amount of water vapor over the northern cap in northern summer. The winter

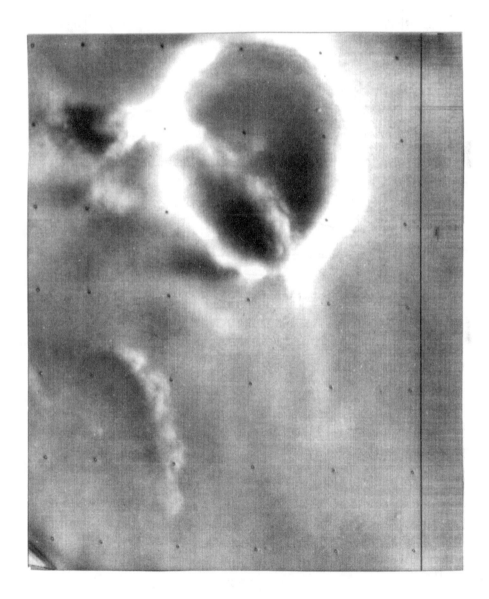

Figure 24. Mariner 9 photograph of a wave cloud over a crater in the Martian north polar region (63° N, 347° W). Courtesy of NASA.

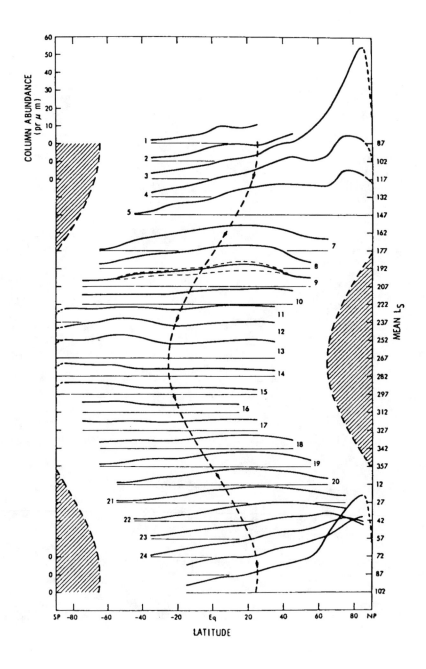

Figure 25. Distribution of water vapor in the atmosphere by latitude and season. Courtesy of Farmer and Doms [19], copyright by The American Geophysical Union.

hemisphere is shown to be essentially devoid of water vapor, but the quantity increased significantly across the equator, its maximum about 100 pr μm at the edge of the north polar cap. The temperature at the cap was 205 K, close to saturation, which indicates the cap was water-ice. Continuing into northern summer, the concentration fell off, becoming evenly distributed by latitude. At the end of northern summer, the vapor distribution was mostly in the northern hemisphere. During fall and midwinter, the distribution was invariant with latitude, yielding about 5 pr μm at all latitudes. At late winter, most of the water vapor moved to the northern hemisphere, and then the cycle repeats itself. Farmer and Doms [19] suggested that there is a yearly net annual transport of water vapor from the south to the north.

One has to consider possible changes in the Martian climate that could affect the stability of ground ice. Figure 26 shows the obliquity of Mars over a 10 m.y. span. This long-term oscillation affects the surface temperature, resulting in deviations from the present mean annual value. These oscillations directly reduce the lifetime of buried ice by periodically increasing the water vapor pressure above the ice layer surface. For instance, an obliquity as low as 10.8° could drop the planet-wide CO_2 atmospheric pressure to approximately 0.1 mbar. "Due to the highly nonlinear dependence of the saturated vapor pressure of water on temperature," Clifford and Hillel [22] point out that "the amount of ground ice which is lost during the high temperature half of the periodic variation will always exceed the amount which is preserved by the low temperature part of the cycle. Therefore, the greater the amplitude of the temperature change, the greater the reduction in ice layer lifetimes."

But the most significant aspect, pointed out by Fanale et al. [21], is the amount of CO_2 adsorbed into the regolith. Due to the climatic fluctuations, as much as 10^{20} g of CO_2 has been driven between the regolith, atmosphere, and polar caps. During this exchange, the CO_2 desorbed from the regolith could act as a carrier gas, flushing the regolith pores of the diffusing water molecules. How this has modified the actual net reduction in the ice layer is not known, but it could considerably shorten the lifetime of ground ice.

5.6.3 Material Conditions

The Viking lander x-ray fluorescence-experiment [52] examined surface material to discover if there were significant amounts of chemically bound water. All of the material gathered revealed a thin fine-grained thickness of eolian particles that is distributed over most of the Mars' surface. The material is unlike anything on earth, and a data fit resulted in iron-rich smectite clays:

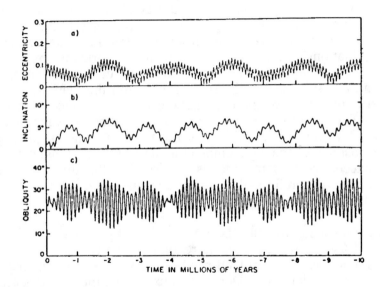

Figure 26. (a) Orbital eccentricity, (b) orbital inclination, and (c) obliquity of Mars for the past 10 million years. Courtesy of W.R. Ward, *J. Geophys. Res.*, [69], copyright by The American Geophysical Union.

(1)   a montmorillonite clay: $Mg_{0.3}Al_{1.70}Si_4O_{10}(OH)_2 \cdot Ca_{0.15}$

(2)   a saponite clay: $Mg_3Al_{0.5}Si_{3.5}O_{10}(OH)_2 \cdot Ca_{0.25}$

(3)   a nontronite clay: $Fe_2Al_{0.5}Si_{3.5}O_{10}(OH)_2 \cdot Ca_{0.25}$

Note the hydroxyl (OH) water in all clays. In addition, considerable amounts of water could be between the layers of crystal structure. Squyres [49] points out that the hydroxyl water could be an important sink of Martian water.

Clifford and Hillel [22] studied the lifetime of a 200 m layer of ground ice buried below 100 m of ice-free regolith for latitudes between ± 30°. Twelve model pore-size distributions that are representative of silt and clay-type soils found on earth simulated various pore structures on Mars. They calculated the flux of escaping water molecules from the buried ground ice layer considering such effects as depth of burial, the Martian geothermal gradient, the regolith porosity, adsorption, surface diffusion, and climatic variation. The material that is presented next follows their paper and is important as it reflects the necessary aspects for studying the behavior of ground ice even though many assumptions they made are severe or questionable.

There is enormous evidence that indicates huge quantities of ground ice may have existed for billions of years. It has been stated earlier that it is impossible to identify with any surety the initial origin and the continual survival of the ground ice. The present atmospheric conditions of temperature and pressure on Mars negate the equatorial regolith from acting as a net annual sink for atmospheric water. "Ground ice which exists in disequilibrium with the atmosphere should experience a net annual depletion, resulting in a preferential transfer of H_2O from the 'hot' equatorial region to the colder latitudes poleward of ± 40°." How then is it possible to explain the existence of ground ice within the equatorial regolith? The explanation given by Clifford and Hillel is that the water found near the equator is a "relic," established very early in Martian history when the atmosphere was warmer. The larger question is: How did this ice manage to survive for about 4 b.y. in disequilibrium with its atmosphere? Earlier it was thought that the depletion could have been retarded by diffusion properties of a shallow layer of fine-grained regolith. Knowing a great deal concerning the physical and chemical properties of the Martian regolith and climate, Clifford and Hillel calculated the lifetime of an unreplenished layer of ground ice lying between ± 30° latitude.

5.6.3.1 The regolith model
The stability of ground ice is governed by the rate at which water molecules can diffuse through the regolith and into the Martian atmosphere. To accomplish this, one must first characterize the physical properties of

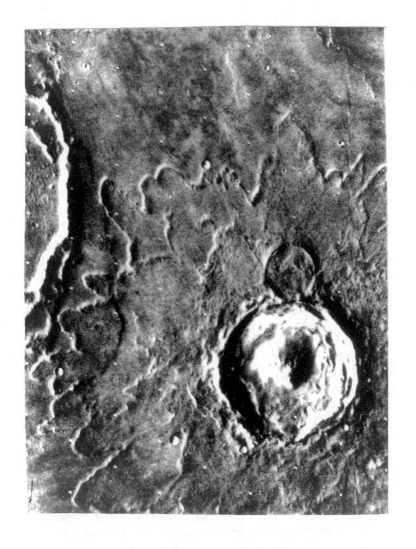

Figure 27. Mariner 9 photograph of the ejecta from an impact on surface of Mars. Ejecta contours suggest a "wet" surface. Courtesy of NASA.

the material medium through which diffusion occurs. This requires knowledge of soil particle size, its specific area, and its porosity. The molecular transport of water involves molecular diffusion, Knudsen diffusion, and surface diffusion. Clifford and Hillel present 12 model-size distributions that are fairly representative of Martian soils.

We begin by addressing how impacts modify the structure of a planetary surface. Figure 27 shows an example of how an impact distorts Mars' surface. The photo shows the "splash" contours of the outer surface of the ejecta. The contours suggest a liquid composition of the Martian surface prior to impact. Note the central cone in the crater. This is typical of a resultant liquid splash, easily duplicated in a laboratory. Obviously upon impact, large quantities of ejecta are produced and dispersed, and intense fracturing of the surroundings and floor take place. Clifford and Hillel (herein referred to as CH) cite references of those who calculated the volume of ejecta and depth of penetration of impact. Estimates of debris averaging 2 km in thickness spread over the planet's surface intermixed with volcanic flows are given. Estimates of porosity at 10 km in depth show the extensive porous nature of the Martian crest.

The soil particle size is different than what one would expect on the Moon due to the presence of an atmosphere and weathering phenomenon on Mars. This would make particle sizes smaller than on the Moon. Experiments conducted by the Viking lander indicate clay size particles with uniform diameter of 0.14 μm.

The diameter d of a soil particle of uniform grain size is given by

$$d = 6/\rho_S S_m \tag{10}$$

where ρ_S is the mean particle density ($\rho_S = 2.65$ g/cm^3 for silicate particles) and S_m is the specific surface area per unit mass ($S_m = 17$ m^2/g from a Viking lander GEX experiment). Figure 28 shows how a soil with a measured specific surface area of 17 m^2/g fine soil could result from both coarse and fine-grain particles. The coarse particles are of $d > 2$ μm and $S_m < 1$ m^2/g. We note in Figure 28 that a soil with $S_m > 100$ m^2/g is confined to less than 17% of the Martian soil by weight if the entire soil is to have a total specific surface area $S_m = 17$ m^2/g.

It is important to note that using measurements of N_2 and CO_2 adsorption to determine S_m are usually much less than values of S_m using measurements of H_2O adsorption. This is due to water molecules, say from the poles, which can penetrate the interlayer clay sheets. The water causes the clay crystal structure to expand, thereby exposing adsorption areas that significantly increase the clay's S_m. Hence, it is presumed the Martian regolith has a broad range of particle sizes and the dominant size is silt or smaller.

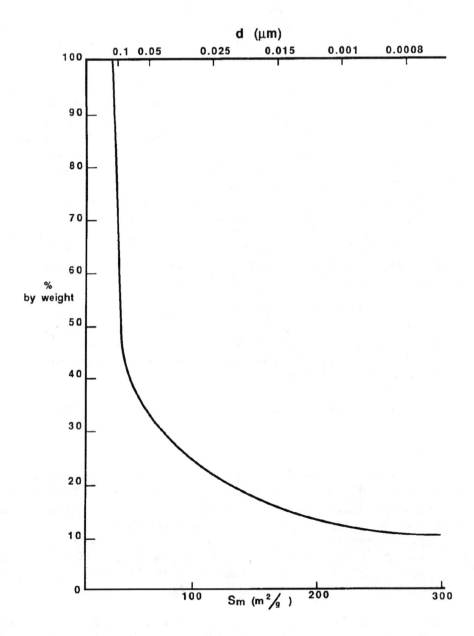

Figure 28. Specific surface area of 17 m²/g fine soil versus content by weight. Courtesy of S.M. Clifford and D. Hillel, *J. Geophys. Res.* [22], copyright by The American Geophysical Union.

Knowing the soil composition, it is possible to estimate its porosity. The effective pore size based on a broad range of particle sizes is customarily determined by the smallest particle size present in the soil. For aggregated soils, the effective pore size is usually determined by the presence of much larger interaggregate pores. The greater the clay content, the greater the likelihood that aggregation is important in determining the soil's structure. To accurately predict the amount of water molecular transport occurring in such soils, it is necessary to construct a system which predicts the broad distribution of pore size.

The nature of most porous solids is that the pore system is modeled by a lognormal distribution $f(r)$ given by

$$f(r) = \frac{1}{\sigma\sqrt{2\pi}} \exp\left[-\frac{1}{2\sigma^2}(\ln r - \ln r_m)^2\right]$$ (11)

where σ is the standard deviation, $f(r)$ is the probability that a pore of radius r occurs about some geometric mean pore size r_m.

As pointed out by CH, no direct measurement of the distribution of pore sizes of the Martian soil was conducted by the Viking lander. A review was made by CH, and they determined that for fine-grained soils having specific surface areas ≥ 17 m^2/g and porosities of 50%, the range of pore sizes should be adequately covered by the 12 pore-size distribution models of Table III. The soils on earth are assumed to possess similar characteristics to the soils on Mars based on their similar physical properties. For lack of a better estimate, CH assumed the regolith had a mean column porosity of 50%. The soils in Table III are all extremely fine grained.

The differential porosity $\Delta\varepsilon(r)$ which occurs in the pore interval $\Delta\ln(r)$ for most of the 12 distributions of Table III is expressed by CH in the general form

$$\Delta\varepsilon(r) = \frac{\Delta\ln(r)}{\sqrt{2\pi}}\left[\frac{E_1}{\sigma_1}\exp\left[-\frac{1}{2\sigma_1^2}(\ln r - \ln r_{m1})^2\right]\right. \left. + \frac{E_2}{\sigma_2}\exp\left[-\frac{1}{2\sigma_2^2}(\ln r - \ln r_{m2})^2\right]\right]$$ (12)

where E_1, r_{m1}, and σ_1 are the micropore porosity, median micropore size, and micropore standard deviation, respectively, and E_2, r_{m2}, and σ_2 are the macropore symbols. Table IV presents the values of these parameters for 12 soil models.

TABLE III. Terrestrial Examples of Pore Size Distribution Models

Pore-Size Distribution	Terrestrial Examples
a	Packed kaolinite clay; fox silt loam, oven-dried.
b	'Grundite," a poorly crystalized illite clay, dispersed, sedimented, and oven-dried. Compact oven-dried.
c	Garfield nontronite, oven-dried; 'Edgar Plastic' kaolinite, dispersed, sedimented, and oven-dried. Compton Beauchamp, a clay soil, chemically treated, and oven-dried.
d	Georgia kaolinite, flocculated, sedimented, and oven-dried with $T = 274°$ C and $p = 216.4$ bars.
e	Hypothetical distribution.
f	'Grundite,' flocculated, sedimented, and oven-dried; and 'Edgar Plastic' kaolinite, dispersed, sedimented, and oven-dried. Compacted 'Grundite,' 19% moisture content.
g	Same as distribution 'c,' but pore sizes have been reduced by a factor of 10. Macon kaolinite, compacted 'Edgar Plastic' kaolinite, and Leda clay, oven-dried with $T = 274°$ C and $p = 216.4$ bars. Drayton soil, a clay soil, chemically treated and oven-dried. Hanslope soil, a clay soil sampled at a depth of 42-74 cm, oven-dried, Ragdale soil—a clay soil sampled at 23-57 cm, oven-dried; Evesham soil—a clay soil sampled at 40-56 cm, oven-dried; and Flint soil, a fine loam soil sampled 65-93 cm, oven-dried.
h	Same as distribution 'a,' but pore sizes have been reduced by a factor of 10.
i	Macon kaolinite, 'Edgar Plastic' kaolinite, flocculated, sedimented, and oven-dried.
j	Clay Spur montmorillonite, oven-dried.
k	Hypothetical distribution.
l	Same distribution as 'a' (reduced by 100) and 'c' (reduced by 10). This pore-size distribution might result from a mixture of 28% illite and 72% kaolinite.

Terrestrial examples were selected on the basis of the resemblance between their published differential or cumulative porosity curves and the pore models. All porosities were normalized to 50% (taken from [22]).

TABLE IV. Model Pore-Size Distribution Parameters [22]

Pore-Size Distribution	Micropores			Macropores		
	Median Pore Size r_{m1}, m	Standard Deviation σ_1	Porosity E_1	Median Pore Size r_{m2}, m	Standard Deviation σ_2	Porosity E_2
a	...	...	...	10^{-6}	1.5	0.5
b	10^{-8}	1.0	0.1	10^{-6}	1.0	0.4
c	10^{-7}	0.6	0.4	10^{-5}	2.0	0.1
d	10^{-7}	0.6	0.3	10^{-6}	2.0	0.2
e	10^{-8}	1.0	0.25	10^{-6}	1.0	0.25
f	10^{-7}	1.0	0.4	10^{-6}	1.5	0.1
g	10^{-8}	0.6	0.4	10^{-6}	2.0	0.1
h	...	...	...	10^{-7}	1.5	0.5
i	10^{-8}	1.0	0.1	10^{-7}	1.0	0.4
j	10^{-8}	0.6	0.4	10^{-7}	2.0	0.1
k	10^{-8}	0.6	0.3	10^{-7}	1.0	0.2
l	...	...	...	10^{-8}	1.5	0.5

Having defined the physical properties characterizing the Martian regolith, we now treat the three models that represent the mechanism for water transport through the soil: ordinary molecular diffusion, Knudsen diffusion, and surface diffusion.

5.6.3.2 Ordinary molecular diffusion

We shall restrict ourselves solely to water vapor transport. The pore size is such that the ratio of the pore radius r to the mean free path λ is large $(r/\lambda > 10)$. Physically, ordinary molecular diffusion is governed by molecular collision, i.e. exchange of momentum. The limit where molecular diffusion begins and Knudsen diffusion ends is not distinct, and thus for pore sizes in the range $0.1 < r/\lambda < 10$, both ordinary molecular and Knudsen diffusion must be considered.

On Mars, $\lambda \sim 8$ μm for water molecules in the atmosphere. Due to the spectrum of pore sizes in the Martian regolith, CH adopted a diffusion model which could predict the molecular flux of water over the full range of gaseous diffusion. The model they chose is called the <u>parallel pore</u> model, due to its simplicity and its success in predicting diffusion and flow through porous medium with typical broad pore-size distribution.

The parallel pore model considers a porous solid as a collection of parallel cylindrical pores of fixed radii. Let A and B represent two components in a binary gas mixture. Let J_A represent the flux of A through component B in a single capillary pore, where the processes are assumed isobaric and isothermal. Thus,

$$J_A = -D_{AB}\frac{dn_A}{dz} + \frac{n_A}{n}(J_A + J_B) \qquad (13)$$

where D_{AB} is the binary molecular diffusion coefficient, n_A is the molecular concentration of component A, n is the total molecular concentration (note: $n = n_A + n_B$), and dz is the elemental length of the capillary pore. The binary molecular diffusion coefficient is given by Wallace and Sagan [15] as a function of pressure p and absolute temperature T as

$$D_{AB} = 0.1654(T/273.15)^{3/2}(1.013 \times 10^6/p) \qquad (14)$$

The flux ratio for a binary gas is due to the ratio of the molecular weights M and is

$$J_B/J_A = -\left(M_A/M_B\right)^{1/2} \qquad (15)$$

Letting

$$\alpha = 1 - \left(M_A/M_B\right)^{1/2}. \qquad (16)$$

CH rewrites Equation 13 as

$$J_A = \left[-D_{AB} / \left(1 - n_A \alpha / n \right) \right] dn_A / dz \tag{17}$$

which can be integrated with respect to z and n_A for steady state as

$$J_A = \left(D_{AB} n / z \alpha \right) \ln \left[\left(1 - \alpha y_{Az} \right) / \left(1 - \alpha y_{Ao} \right) \right] \tag{18}$$

where

$$y_{Ao} = n_{Ao} / n, \quad \text{at } z = 0 \tag{19}$$

$$y_{Az} = n_{Az} / n, \quad \text{at } z = Z \tag{20}$$

$$n = p / kT \tag{21}$$

and k is the Boltzmann constant.

5.6.3.3 Knudsen diffusion

Knudsen diffusion is the dominant diffusion mechanism when molecular transport occurs in very small pores ($r/\lambda < 0.1$). Here collisions between diffusing molecules and the pore walls greatly exceed any other collisions with other molecules. For Knudsen diffusion, we do not use Equation 18 for the flux, but a similar form. CH obtains an expression for the flux that enables them to predict the gas phase transport through almost any size pore. Dividing the right-hand side of Equation 13 by $(1 + D_{Ao}/D_{kA})$, where D_{kA} is the Knudsen diffusion coefficient of component A

$$D_{kA} = \frac{2}{3} r \left[\frac{8 \, TR}{\pi \, M_A} \right]^{1/2} \tag{22}$$

for a straight cylindrical pore of radius r. CH obtain the gaseous flux of component A for any small pore size as

$$J_A = \frac{D_{AB} p}{k \, T z \, \alpha} \ln \left[\frac{1 - \alpha y_{Az} + D_{AB} / D_{kA}}{1 - \alpha y_{Ao} + D_{AB} / D_{kA}} \right] \tag{23}$$

This expression has been verified experimentally by four separate investigators in the three regions of ordinary molecular diffusion ($r > 80 \ \mu$m) intermediate diffusion ($0.8 \ \mu$m $< r < 80 \ \mu$m), and Knudsen diffusion ($r < 0.8 \ \mu$m) [22].

All the above holds for a single capillary pore. We need to apply Equation 23 to a porous solid with its broad spectrum of pore diameters, which can be easily handled by simply summing over the pore range:

$$J_A = \frac{D_{AB}\, p}{q\, k\, T\, z\, \alpha} \sum_{r_{min}}^{r_{max}} \ln\left[\frac{1 - \alpha y_{Az} - D_{AB}/D_{kA}}{1 - \alpha y_{Ao} + D_{AB}/D_{kA}}\right] \Delta\varepsilon(r) \qquad (24)$$

where it is recalled $\Delta\varepsilon(r)$ is given by Equation 12 and represents the porosity in the pore interval Δr, and q allows for dead-end pore space and is called the tortuosity factor. It has a value of approximately 5 for a porosity of 50%.

5.6.3.4 *Results and analysis*

In order to calculate the lifetime of an unreplenished layer of ground ice between ± 30° latitude, several assumptions are required: CH assumed i) the amount of ice contained in a vertical column was 10^4 q/cm^2, ii) the ice occupied the available pore space in a single layer of a depth 100-300 m, iii) the upper 300 m was isothermal at a temperature equal to the mean annual surface temperature 210 K $\leq T \leq$ 230 K, and iv) the equilibrium vapor pressure of water just above the surface of the buried ice layer was

$$p = 1.333 \times 10^3 \exp(-5631.1206/T + 18.95304\log(T)$$
$$-3.861574T + 2.77494 \times 10^{-5} T^2 - 15.55896) \qquad (25)$$

where p is in units of dynes/cm^2. Equation 25 is plotted in Figure 29.

The flux J of escaping water molecules from the buried ice on Mars was calculated by CH using Equation 24 for each of the 12 pore-size distributions of Table IV, using the different porosity $\Delta\varepsilon(r)$ of Equation 12 and the value of the parameters also given in Table IV. For the 50 g/cm^2 of water contained in each 1 m thickness of regolith, the lifetime $d\tau$ was calculated using

$$d\tau = \frac{50\, N_0}{J_{H_2O}\, M_{H_2O}} \qquad (26)$$

where J_{H_2O} is the water molecular flux, N_0 is Avagadro's number, and M_{H_2O} is the molecular weight of water. The total lifetimes of the buried ice layer are summarized in Table V by CH. Examining Table V, we see variations from 166 m.y. to 133 b.y., which is quite a spread. Taking the average of the annual mean surface temperatures of 220 K, only 4 pore size will result in a lifetime of ice for over 4 b.y. As pointed out by CH,

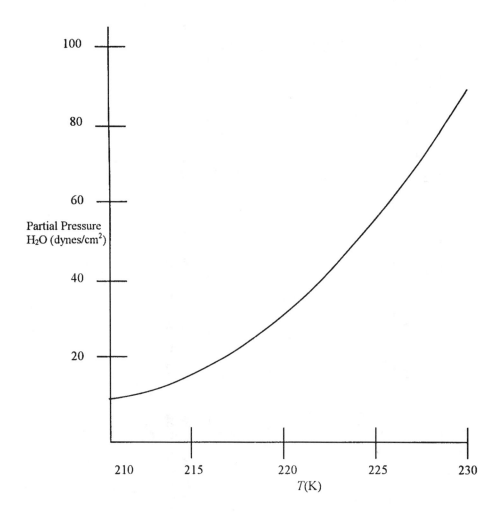

Figure 29. The saturated vapor pressure of H_2O as a function of temperature. Redrawn from S.M. Clifford and D. Hillel, *J. Geophys. Res.* [22], copyright by The American Geophysical Union.

TABLE V. Lifetimes of a Buried Ground Ice Layer

Pore-Size Distribution	Mean Surface Temperature K								
	210	212.5	215	217.5	220	222.5	225	227.5	230
a	2.6	1.7	1.1	0.8	0.5	0.4	0.3	0.2	0.1
b	3.8	2.5	1.7	1.2	0.8	0.6	0.4	0.3	0.2
c	4.3	2.8	1.9	1.3	0.9	0.6	0.5	0.3	0.2
d	5.1	3.3	2.2	1.5	1.1	0.8	0.5	0.4	0.3
e	6.1	4.0	2.7	1.9	1.3	0.9	0.7	0.5	0.3
f	8.8	5.8	3.9	2.7	1.9	1.4	1.0	0.7	0.5
g	11	7.2	4.9	3.4	2.4	1.7	1.2	0.9	0.6
h	14	9.1	6.2	4.3	3.0	2.2	1.6	1.2	0.8
i	26	17.4	12	8.3	5.9	4.2	3.1	2.3	1.7
j*	39	26.0	18	12	8.8	6.3	4.6	3.4	2.5
k*	48.3	32	22	15	11	7.8	5.7	4.2	3.1
l*	113	75	51	36	26	19	14	10	7.4

In billions of years.

*Pore-size distributions that are indicative of clay soils which display little or no large-scale aggregation and have specific surface areas that substantially exceed the figure inferred for the Martian regolith (taken from [22]).

the calculated data show that low temperatures and small pore size enhance the ability of ground ice to exist in disequilibrium with the Martian atmosphere. Note that a temperature increase of only a few degrees reduces the lifetime of ground ice by approximately 25%. This, in turn, is a reflection on the partial pressure of water in equilibrium with ice: namely, 7 dynes/cm^2 at $T = 210$ K to 90 dynes/cm^2 at $T = 230$ K. For the temperature range characteristic between $\pm 30°$ latitudes, i.e. 215-225 K, there is an increase of 300% in the saturated vapor pressure of water over an ice layer surface. This enormous increase can cause a 70% drop in the ice layer lifetime between the coldest and warmest latitudes. Hence for the warmer latitudes (mean annual temperature > 217 K), Table V shows that a regolith with 50% porosity and surface area of 17 m^2/g could maintain a 200 m thick layer of ice buried 100 m below an ice-free surface for as long as 4 b.y. Obviously, changing the regolith temperature and/or changing the depth of the buried ice will yield different results. CH also assumed an isothermal process, and it is known that even a small temperature gradient can have a marked effect on ice stability. The same can be said of the porosity. On earth, the porosity of the crust declines with depth due to compaction, cementation, and igneous reconsolidation. On Mars, it does not seem unreasonable this would be the case for the first few km of regolith.

5.6.3.5 *Surface diffusion*
It is interesting to note there is no totally correct model of surface diffusion, and thus it is difficult to determine how important this diffusion role is to the transport of water through the regolith. CH make an order of magnitude assessment. They use a Fick's law relation for the surface flux J_S per unit cross-section of regolith.

$$J_S = -\frac{D_S}{q_S}\rho_S S_m \frac{dc_S}{dz} \qquad (27)$$

where D_S is the surface diffusion coefficient of an adsorbate

$$D_S = 1.6 \times 10^{-2} \exp(-0.45 Q/mRT) \qquad (28)$$

and q_S is the surface tortuosity ($q_S = 5.0$ for a gas). The particle density is $\rho_S = 2.65$ g/cm^3, S_m is the specific surface area of the regolith ($S_m = 17$ m^2/g), c_S is the surface concentration of H$_2$O molecules ($c_S = 6.3 \times 10^{14}$ molecules/cm^2), Q is the heat of adsorption (kcal/mol), and m is an adjustable parameter ($m = 1, 2, 3$) that depends on the nature of the adsorbant and the surface bond. For example, for a polar adsorbate bonded by Van der Waals' and electrostatic forces to an insulating material (such as clay), $m = 1$. Knowing Q, we can obtain a fairly reliable estimate of the surface diffusion coefficient D_S of water adsorbed on any mineral surface. CH find that the surface flux of water molecules from their model ice layer is approximately 6.2×10^8 molecules/cm^2·s. Though this flux results in a near negligible change in the results presented in Table IV, it does set a limit of a maximum of 10 b.y. for the ice layer. On the other hand, if the specific surface area is 200 m^2/g, then J_S could result in a value over 10^{10} molecules/ cm^2·s resulting in a much smaller maximum ice layer lifetime between 10 m.y. and 1 b.y.

5.7. CONCLUSION

Though the CH model is extremely interesting and of immense value, the parameters lack experimental justification. There is simply too much we do not know. For example, (1) the structure of the Martian regolith may be dissimilar to terrestrial clay type soils; (2) most of the fine-grained terrestrial soils in the distribution considered by CH could not preserve a 200 m thick layer of ground ice buried beneath a 100 m ice-free surface for as long as 4 b.y.; (3) a soil's gaseous permeability cannot be determined by particle size and porosity alone; (4) existence of a temperature gradient can nullify all CH results, giving an even greater reduced lifetime; and (5) climatic changes in the CO_2 atmosphere could accelerate the evaporative loss of Martian ground ice. We are still confronted with the huge morphological evidence that enormous quantities of ground ice currently exist between the $\pm 30°$ latitudes. CH suggest the following answers:

(1) the ground ice inventory may be greater than the 10^4 g/cm^2 assumed in the model,

(2) the geothermal gradient on Mars may be extremely small, and

(3) the ground ice may have been replenished.

6

ALTERNATE SOURCE OF WATER DEPOSIT

The last area for a deposit of water to be discussed is in the polar cap. This has been mentioned briefly earlier, but now we need to explore it in more detail. Squyres [49] presents an excellent overview which is capsulized below. There are principal regions of the polar cap that need to be addressed: (a) layered deposits, (b) perennial ice zone, and (c) seasonal frost cap.

6.1 LAYERED DEPOSIT ZONE

Layered deposits are found at both north and south poles and cover a region of 80°-85° latitude. Squyres calls the zone a "thick sequence of thin, horizontal layers," ranging in thickness from 10 to 50 m. The individual layers run continuously laterally for hundreds of kilometers, and though the total thickness is difficult to calculate, it could be 1-2 km in the south cap and 4-6 km in the north cap (Dzurisin and Blasius [53]).

The layered deposits are relatively recent since the zone is free of impact craters. The zone consists of a "roughly spiral pattern of deep accurate troughs that expose the edges of the layers on their walls." The trough slopes are about 1°-8°, where the troughs are 10-20 km wide.

On top of the layered deposits lies perennial ice. This is ice that exists throughout the Martian year. It is not as extensive as the layered deposits, reaching almost to the extent of layered deposits at the north pole, but reaching a much smaller area at the south pole. At the north pole, the typical surface temperature is 205 K, which is higher than the 148 K necessary for the saturation temperature of CO_2 (considering a mean surface pressure of 6.1 mbar). Thus, the northern perennial ice is likely water. The albedo of ice is approximately 43%, which means there is an admixture of dust or some such similar dark debris. Kieffer [54] presents results from Viking observations that the south pole is considerably more complicated than the north pole, that late summer temperatures were colder and the CO_2 frost existed at the surface. This cold temperature may be due to screening of solar radiation by global dust storms that occur in the late summer. Figure 30 is a portion of the northern polar cap edge showing outliners of ice resting on a mesa of layered terrain about 80 km wide. "Slopes of uniform width and declivity facing outward from the center of the residual cap defrost earlier than level areas because of their inclination."

Another fascinating photograph showing the underlying layered terrain revealed on the slopes facing away from the south pole is Figure 31. The photograph was taken when the seasonal frost cap had retreated to its minimum area. Whether water is present in the southern cap or not is still uncertain, since cold summer temperatures prevent detection of water. It is probable that the southern cap acts as a trap for both CO_2 and H_2O. If water is trapped, it cannot get out because temperature never reaches frost point. The southern cap is accumulating CO_2 and is growing in size at the expense of the northern cap. In 25,000 years, conditions will be reversed.

In Figure 31 the valleys and escarpments are approximately equally spaced and about 50 km apart. Removal of frost on the equator-facing slopes gives the pole its characteristic summer swirl texture. "The walls and escarpments range in height from 100 m to 1000 m with slopes as high as 6°" (Carr [3]). The layering is visible largely due to albedo contrasts between individual layers. "Because of the small number of craters, the layering is considered recent and not due to some self-annealing process or infilling" (Carr [3]). One also notes the accumulation of dust. "The combination of eccentricity variations (0.14 to 0.01 over 2×10^6 years) and precession (less than 0.6 with a period of 95,000 years) causes periodic variation of dust deposits on time scales up to 2×10^6 years" (Carr).

A number of physical models have been proposed to explain the layered deposits. Cutts [55] suggested that the layered deposits were formed early in Martian history and presently are experiencing eolian erosion-forming troughs. Figure 32 shows the layers appearing as giant mesas. The south pole layered terrain is one of the most striking Martian surface features. From the ground the layers might look like the masses of the U.S. Southwest. Individual layers are probably 20-50 m thick. Their origin is a mystery. Smooth, gracefully sculptured surfaces with gentle

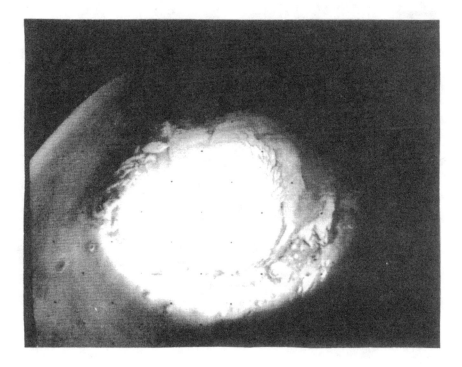

Figure 30. Mariner 9 photograph of Mars' north pole. Courtesy of NASA.

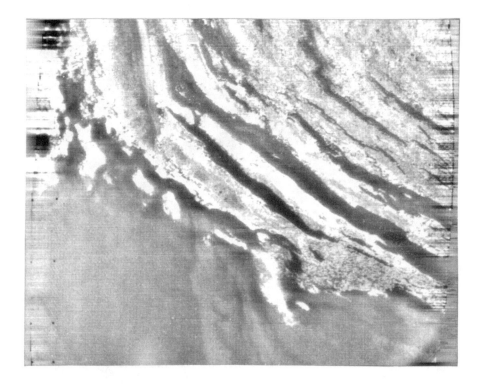

Figure 31. The residual south polar cap, seen resting on a mesa of layered terrain. Patchy ice mass occurs
due to disconnected remnants. Courtesy of Mariner 9, NASA.

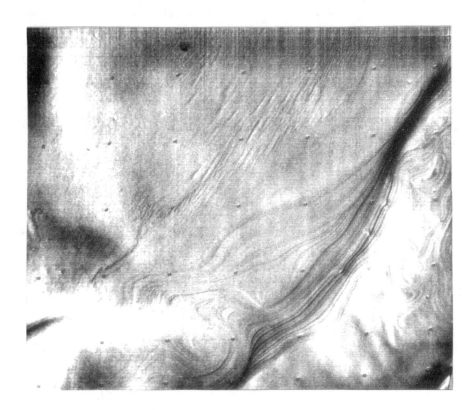

Figure 32. Mariner 9 photograph of Mars' south pole layered terrain. Courtesy of NASA.

slopes are characteristic of this terrain.   The upper edges, unlike those of slopes in the pitted plains, are rounded.   Layered terrain is essentially crater form, indicating that it is relatively young or recent erosion.   It is believed that the seasonal frost cap helps in the formation of the layered deposits, trapping particulates as the ice forms.   The layered deposits could conceivably contain 10^3 g cm^{-2} equivalent water and may be the largest single sources of water on Mars.   Three additional photographs of the south polar cap show details of the unusual topography of the cap. Working our way up to the pole, Figure 33 shows outliers of ice resting on a mesa of layered terrain about 80 km wide.   The slopes are of near uniform width, and declivity facing outward from the center of the cap defrost earlier than level areas due to this sloping.

Figure 34 shows the strange patterns of frosted layered terrain distorting the topography.   Note the fine layering on the partly defrosted slopes and how similar it looks to gigantic terrestrial glacier fields.

Figure 35 was taken late in the space mission when the southern cap had reached its limit of retreat.   The underlying layered terrain is seen as gentle slopes facing away from the pole.

The top layer of the polar cap ice is identified as the seasonal frost cap. During each Martian winter, the layered deposits and perennial ice are covered with a seasonal frost cap.   Figures 36 and 37 show the frost cap in the north polar region.   Figure 36 shows the cap covering about 2700 km wide after the vernal equinox.   The cap has a near circular shape.

Figure 37 shows the cap as being very near its minimal condition.   The cap is about 1000 km across, and its curved patterns are viewed as a collection of stacked plates.   Each plate may have 20-40 separate layers, with an aggregate thickness of 1 km.   The permanent cap is probably mostly water, because a permanent mass of frozen CO_2 would collect water even from Mars' dry atmosphere.   The seasonal frost, however, is probably a thin layer of solid CO_2 condensed from the atmosphere, with frost temperatures ~150 K (Neugebauer [56]).

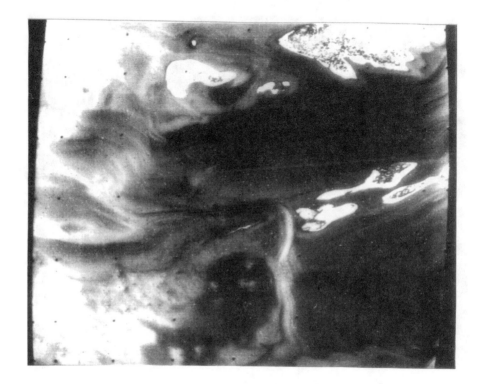

Figure 33. Mariner 9 photograph of Mars' southern polar cap edge showing outliers of ice resting on a mesa of layered terrain (83° S, 37° W). Courtesy of NASA.

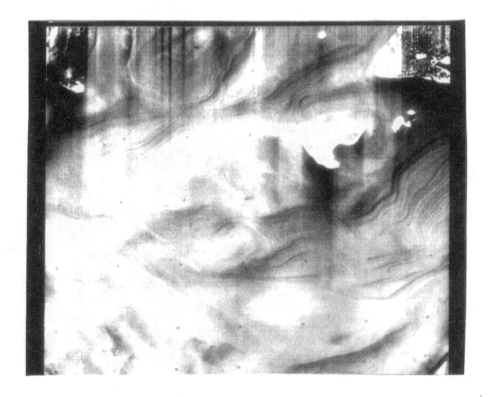

Figure 34. Mariner 9 photograph showing details of Mars' southern polar cap's partly frosted layered terrain (84° S, 25° W). Courtesy of NASA.

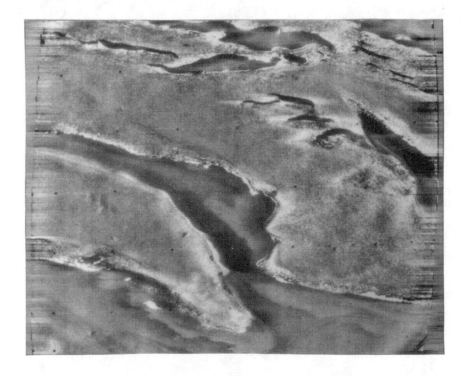

Figure 35. Mariner 9 photograph showing the underlying layered terrain as gentle slopes facing away from the pole (86° S, 97° W). Courtesy of NASA.

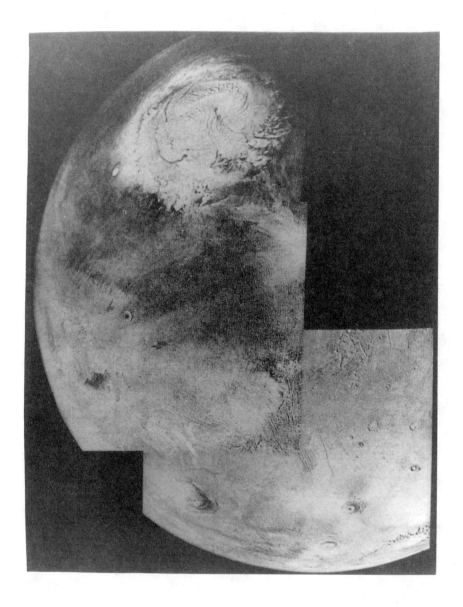

Figure 36. A mosaic of the Mars polar cap. From Mariner 9. Courtesy of NASA.

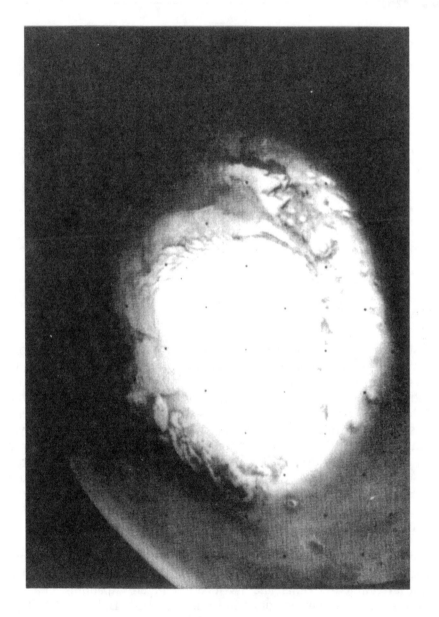

Figure 37. Mariner 9 photograph of the frost cap of Mars' north polar region. Courtesy of NASA.

7

HYDRODYNAMICS OF CATASTROPHIC FLOODING ON MARS

7.1 BACKGROUND

At this stage of the discussion we are faced with a tautological situation: it appears highly presumptuous to speculate on catastrophic flooding (a mechanism that may have created some of the channels on Mars) before we identify the source of the responsible fluid. We cannot prove that what caused the creation of the outflow channels was water. It could have been wind, viscous earth-like flows, glaciers, lava, or combinations of all five, but all that we have stated in the previous chapters points out that water is the prime suspect. Milton [57] states that certain features of the large Martian channels could only have resulted from running water. Milton concludes that Mars experienced a fluvial experience sometime in its history. He points out that the "braided reach" of the Mangala Vallis is the strongest argument for liquid water on Mars.

Figure 38 shows the cross-cutting relationships of secondary scouring channels. Due to the immense scale of the fluctuating discharge regimen of terrestrial braided streams and its location in an expanding channel reaching immediately downstream from a constriction, Baker and Milton [58] suggest that it is more likely an expansion fan that developed during catastrophic flood discharging and that the residual surface of deposits was subsequently scoured by either secondary flows or subsequent flooding.

Figure 38. The cross-cutting relationships of secondary scour channels. Note the expansion bar developed during catastrophic flooding. Mariner 9 photograph. Courtesy of NASA.

Milton [57] stipulated that this braided pattern could be produced only by a high density, low viscosity liquid moving rapidly over a particulate bed. Such a requirement excludes wind, lava flows, and ash flows. It is not necessary to identify other theories as the arguments for water are just too strong. Baker [59] summarizes various fluid flow processes with specific morphological features of the outflow channels on Mars in Table VI. Note, every fluid flow system can explain a specific aspect of the outflow channel morphology (marked by an X in the table), however some features cannot be explained (–) or are questionable (?). Only catastrophic flooding has the ability to explain all features, followed by glacier ice. Thus, only catastrophic flooding appears capable of producing all the morphological features of the Martian outflow channels. Again, this is not a proof that water created the channels.

We shall now address the magnitude or scale of the flooding.

The fluid dynamics of Martian catastrophic flooding is difficult due to the unknown properties and channel geometries involved. All one can do is make judicial estimates that will give orders of magnitude and hence are subject to large error.

We have noted, using Table II, that liquid water can either freeze or evaporate, depending upon pressure and temperature. As stated earlier, if we assume a denser atmosphere than present, then liquid water could flow under the ice over the channels, effectively insulating the water from the cold surface temperature. That water could have large concentrations of dissolved salts from ground material, thereby lowering the freezing point and raising the boiling point to keep the fluid liquid. The fluid dynamic phenomenon we investigate is open channel flow, and for terrestrial behavior, any text in fluid mechanics (e.g. Granger [60]) presents the dynamics of the fluid flow. The appropriate formula to use for Martian open channel flow is (Carr [3])

$$\tilde{V} = \frac{0.5}{n} D^{2/3} S^{1/2} \tag{29}$$

where $\tilde{V}$ is the average flow velocity (m/s), D is the flow depth (m), S is the energy slope, n is the Manning roughness coefficient, and 0.5 approximates adjustment for the Martian gravity. Table VII summarizes the flow parameters for the Martian outflow channels and two terrestrial analogs (the Mississippi River and the Missoula Floods). One notes the similarity between the Missoula Floods and the Ares Vallis discharge rates, but dissimilarity with the Mississippi River flooding. Any shooting flow occurs at abrupt flow constrictions, wherein maximum erosion takes place.

TABLE VI. Features of Martian Outflow Channels (Source: From THE CHANNELS OF MARS by Victor R. Baker. Copyright © 1982. By permission of the University of Texas Press.)

Morphological Features	Wind	Mudflow	Glacier	Lava	Catastrophic Flood
Anastomosis	?	X	X	X	X
Streamlined uplands	X	X	X	?	X
Longitudinal grooves	X	X	X	?	X
Scour marks	X	?	?	?	X
Scabland	?	–	?	–	X
Inner channels	?	X	?	X	X
Lack of solidified fluid at channel mouth	X	–	X	–	X
Localized source region	?	X	X	X	X
Flow for thousands of kilometers	X	–	X	X	X
Bar-like bed forms	?	?	?	?	X
Pronounced upper limit to fluid erosion	–	X	X	X	X
Consistent downhill fluid flow	?	X	X	X	X
Sinuous channels	?	X	X	X	X
High width-depth ratio	X	X	X	–	X
Headcuts	–	?	X	–	X

TABLE VII. Properties of Martian and earth Channels (Source: From THE CHANNELS OF MARS by Victor R. Baker. Copyright © 1982. By permission of the University of Texas Press.)

Channel	Depth D (m)	Slope S	Channel width W (m)	Mean Velocity $\tilde{V}$ (m/s)	Discharge (m^3/s)
Maja Vallis	100	0.02	8×10^4	38	3×10^8
	10	0.02	5×10^4	8	4×10^6
Ares Vallis	100	0.001	2.5×10^4	27	2.1×10^7
	10	0.001	2.5×10^4	6	0.45×10^6
Mangala Vallis	100	0.003	1.4×10^4	15	2×10^7
	10	0.003	1.4×10^4	3	4×10^5
Mississippi River	12	0.00005	8.3×10^2	3	3×10^5
Missoula Floods	150	0.001	7×10^3	16	2×10^7
	60	0.002	2.4×10^3	16	2.6×10^6

There is ample evidence where flooding eroded wrinkle ridges (e.g. in Kasei Vallis and in western Chryse Plantia of Mars) and there was erosive stripping of channel floors (e.g. Maja and Ares Vallis).

Erosion is a complex phenomenon. It involves vortex action, stream-lining, and cavitation. Catastrophic flooding is highly likely for shaping the Martian outflow channels through the interaction of the above three fluid dynamics phenomenon. We shall treat each separately.

7.2 CAVITATION EROSION

Cavitation erosion in open channel flow is localized at three primary points: (a) at expansions where the adverse pressure gradient results in the collapse of cavitation bubbles that had formed at upstream conditions (where velocities were large). The physics is quite simple to understand. At convergent sections where velocity of the liquid increases, the pressure decreases. If the velocity is large enough or constriction is small enough, the pressure can reach vapor pressure, such that minute bubbles form. These bubbles must be attached to a nucleus in order to grow, usually mi-nute particulates being carried downstream by the flood waters. When the bubbles reach a zone of high pressure (or low velocities) such as in the expansion zone of the channel, the bubbles collapse. They collapse in a unique way. Considering a bubble is like a filled balloon, imagine poking your finger at the balloon surface toward the center of the balloon. A jet of water quickly forms and impacts on the nuclei (see Figure 39) with enormous pressures on the order of 2×10^9 N/m^2, on an area microscopi-cally small, smaller than the diameter of the bubble. Such pressures can shatter the minute surface layer of rocks that are in contact with the col-lapsing bubble. The fractured surface layers are quickly swept away by the flood water until the process repeats itself. This pitting action is called cavitation erosion. (b) Zones of flow separation due to discontinuities in the rocks can form vortex filaments and tubes of swirling flow whose ve-locity can be large enough to form cavitation. (c) High concentrations of vorticity due to localized swirl can also cause cavitation.

The critical mean stream velocity permitting terrestrial cavitation V_e (m/s) comes from Bernoulli's equation:

$$V_e = \sqrt{\frac{2}{3}g}\sqrt{\left(\frac{p_a - p_v}{\gamma}\right) + D} \qquad (30)$$

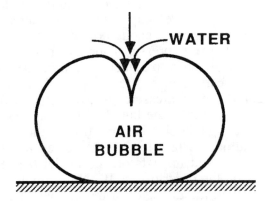

Figure 39. Collapse of an air bubble.

where the $(p_a - p_v)$ term is negligible for the present Martian environment and D is the stream depth (m). A set of representative curves of the critical conditions for cavitation in flood flows on Mars and earth for various atmospheric pressures is presented in Figure 40. For the present Martian atmosphere, Baker [59] obtains

$$V_e = 1.6\sqrt{D}. \tag{31}$$

Martian cavitation bubble pressures become important for bedrock erosion at depths $D \geq 30$ m. The relevant portion of Figure 40 is the right-hand side, not the left, where critical conditions become independent of atmospheric conditions and the above equation applies. It is just possible that the flood velocities were at critical mean stream velocities such that for deep depths, cavitation could have been possible. It is interesting to note that if ice covered sections of the outflow channels, such as at entrances to constrictions, then this could help explain why the outflow channels are restricted to very large features. At the smaller scale, water would cavitate out.

An alternate way to obtain the velocity of flood waters is to use empirical results from terrestrial flooding. For example, Costa [61] obtains

$$V_e = 57\, D^{0.46}, \; 1 \, cm \; \leq D \leq 500 \, cm \tag{32}$$

as the critical mean stream velocity permitting cavitation, where D in the above equation is the diameter of the maximum size gravel or boulders transported (where D is in cm and V is cm/s).

7.3 STREAMLINING

One of the strongest pieces of evidence for fluid flow is the effect of streamlining around fixed objects found in outflow channels. Figure 41 shows several streamlined shapes that could only have been produced by the transport of a fluid (be it liquid or gas). The photo is of the plains of Chryse. Flow from Maja Vallis diverged across the plains and deeply scoured the area. Ridges that are similar to those found on lunar maria partly obstructed the flow which was funneled through gaps and low points where intense erosion occurred. The view is 155 km across.

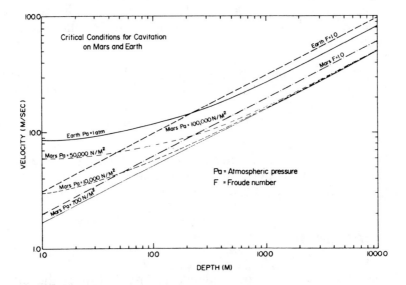

Figure 40. Critical conditions for cavitation in Martian and earth floods, for various atmospheric pressures. From V. Baker, *J. Geophys. Res.* [4], copyrighted by American Geophysical Union.

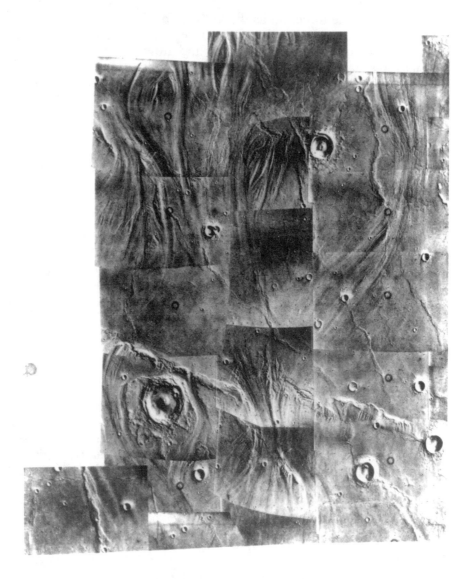

Figure 41. Mariner 9 photograph of the Martian plains of Chryse showing streamlined shapes of islands produced by transport of a fluid. The view is 155 km across. Courtesy of NASA.

The very act of streamlining involves the fluid seeking a path that offers the least resistance to the flow, and the geometry of the streamlined shape gives an excellent illustration of the pattern of the streamlines of the flow, as illustrated in Figure 42. The streamlines represent a line whose tangent is in the same direction as the velocity of the flow. In Figure 42, the zero streamline, represented by $\psi = 0$, represents the optimum shape for minimum resistance (drag), since a fluid flow will always follow a path of least resistance. Denoting D_T as the total drag force on one of the streamlined shapes, one can show (c.f. Granger [60]) that it is composed of pressure drag D_p, friction drag D_f, and wave drag D_w, such that

$$D_T = D_p + D_f + D_w \qquad (33)$$

The pressure drag, D_p, is due to that component of the pressure in the direction of the drag (i.e. the velocity), or

$$D_p = -\int_S (p\cos\phi)_o \, dS \qquad (34)$$

where S is the total surface area, ϕ is the angle between the normal to the surface element and the absolute flow direction, and the subscript zero denotes the pressure along the $\psi = 0$ streamline. The friction drag, D_f, is simply the sum of the components of the shear stresses on the $\psi = 0$ streamline in the direction of the velocity:

$$D_f = \int_S \tau_o \sin\phi \, dS \qquad (35)$$

where τ_o is the shear stress on the $\psi = 0$ streamline. The wave drag is important if surface waves from the flood contact the obstacle in its path. For such obstacles as shown in Figure 41, the wave drag can be neglected. Thus, knowing the $\psi = 0$ shape, the pressure and friction drag can be calculated using Equations (34) and (35) such that we can determine dimensionless drag coefficients:

$$C_{D_T} = D_T \big/ \left(1/2\,\rho V^2\right) A_p \qquad (36)$$

$$C_{D_p} = D_p \big/ \left(1/2\,\rho V^2\right) A_p \qquad (37)$$

$$C_{D_f} = D_f \big/ \left(1/2\,\rho V^2\right) S \qquad (38)$$

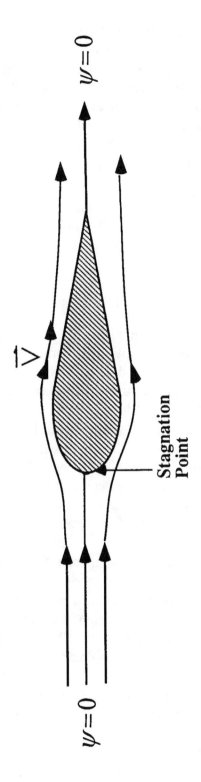

Figure 42. Ideal fluid flow past a circular obstacle.

where $(1/2\rho V^2)$ is the dynamic pressure, A_p is the planar or projected area (the area of the wetted section orthogonal to the flow) and S is the wetted or surface area. Granger [60] shows that the drag coefficients are functions of density, ρ, dynamic viscosity μ, flow velocity V, a characteristic reference length L, and surface roughness ε. These fluid parameters can be combined in dimensionless fashion such that

$$C_D = C_D(R_L,\ \varepsilon/L) \tag{39}$$

where R_L is the Reynolds number

$$R_L = V_e L/\nu \tag{40}$$

where

$$\nu = \mu/\rho \tag{41}$$

A plot of C_D versus R_L is shown in Figure 43 for a variety of shapes. Note, as the width W approaches zero we approximate a flat plate that lies in the plane of the flow. A flat plate's drag is solely friction drag, pressure drag being zero. In matching the Martian streamlined shapes to the shapes in Figure 43, Baker [59] showed the best fit was the $L = 3W$ strut, where $L = 2\sqrt{A}$, A being the area of the Martian streamlined object. Without knowing what the drag on the body is, it is impossible to calculate the velocity of the flow. However, if we know the depth D of the channel, the width W, and the flow rate Q, we can calculate the average velocity. For example, from Table VII, for the Maja Vallis depth of 100 m, channel width 8×10^4 m, and flow rate 3×10^8 m^3/s, an average velocity of 37.5 m/s is achieved. Using Equation 32, the diameter of a boulder transported by flooding becomes 91 m. Using the lower scale for the Maja Vallis, $D = 10$ m, $W = 5 \times 10^4$ m and $Q = 4 \times 10^6$ m^3/s, one obtains $V_e = 8$ m/s, or a boulder diameter of 3 m. Figure 44 is a photo of basalt entablature measuring 18 m $\times$ 11 m $\times$ 8 m that was transported in a Pleistocene flooding on earth. This boulder is located on Ephrata Fan, 2.5 km west of the Rocky Ford Creek Fish Hatchery, Ephrata, Washington. Such large velocities over huge boulders could transport considerable debris, creating enormous outflow channels on Mars. It is clear that substantial flows and erosion are possible in a cold climate under a protective ice cover and that the Martian surface area near the end of heavy bombardment was an ever-changing event with frequent burials of material by the ejecta of large impacts. Also, all photographs from Mariner 9 and Viking lander show

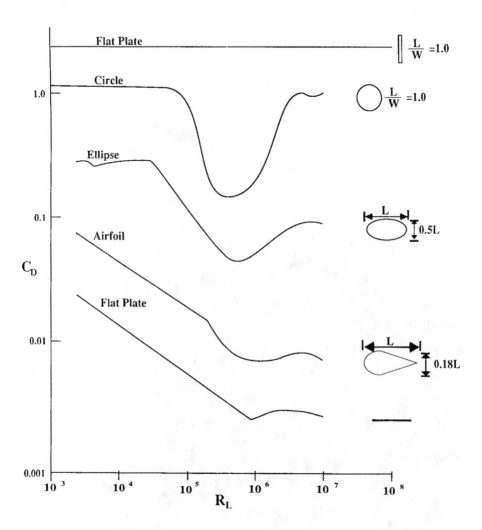

Figure 43. Drag coefficient versus Reynolds number for a variety of land shape forms.
Source: R.A. Granger [60].

Figure 44. Photograph of basalt entablature measuring 18 m × 11 m × 8 m that was transported during a Pleistocene flooding on earth. From THE CHANNELS OF MARS by Victor R. Baker, Copyright © 1982. By permission of the University of Texas Press.

little evidence of run-off erosion on the intervalley system. This strongly suggests precipitation was <u>not</u> responsible for the carving of the channels. Rather, the landforms are more typical of those formed by ground water sapping: i.e. the gradual removal and flow of subsurface fluid. Due to their immense size, the Martian channels bear only the most superficial resemblance to terrestrial landscape. Figure 45 is a satellite photograph of a high plateau across the Sahara Desert. Note the dendritic pattern similar to the dendritic pattern on Mars in Figure 7. Only those in terrestrial regions caused by catastrophic flooding is there a resemblance. Then, too, the transported sediment could have settled and been distributed or covered by eolian transport. The depths of the channels may thus be totally misleading. The mechanism that causes transport of material as well as causes scouring is the third form of erosion: vortex action.

7.4 VORTEX ACTION

There are two significant types of vortex action in catastrophic flooding: longitudinal and vertical vortex phenomena. Consider first longitudinal vortex phenomenon. Figure 46 shows the classic case of "shear flow in a corner." The blunt body could represent a rock, boulder, or outcropping. When a fluid moves past a bed or floor, the viscous stresses in the flow due to viscosity and a velocity gradient du/dz result in a nonlinear velocity distribution with the boundary condition of zero slip at the bed, $z = 0$. As the flow nears the blunt body, an upper lamina of fluid deforms downward and then outward, resulting in a swirl or vortex tube that bends around the blunt body in the shape of a horseshoe. A different lamina will follow a similar behavior, creating a vortex tube inside or outside the former tube depending upon whether the lamina is below or above the previous lamina. Figure 47a is a photograph of the horseshoe vortex created in the laboratory, and Figure 47b is a schematic of the vortex identifying the salient regions of the flow. It is important to point out the breakdown of the vortex which occurs downstream of the maximum thickness of the blunt body (in the region of divergent flow). The primary vortex breaks down into two swirling helical vortices, whose swirling motion is not unlike that of beaters on a mixing machine. The forward part of the vortex breakdown is like a hemispherical cap of near relative stagnation, and the entire vortex breakdown arrangement looks like an octopus with two swirling tentacles. The beating action of the secondary vortices scours the river bed leaving large holes. Figure 48 is an example of such scouring near the 18 m × 11 m × 8 m boulder of Figure 44 in Ephrata, Washington. Note the streamlining of the various sediments transported and collected in shapes that are very similar to shapes found on Mars.

Figure 45. Satellite photograph of a high plateau across the Sahara Desert showing dendritic patterns similar to the dendritic patterns on Mars in Figure 7. Courtesy of NASA.

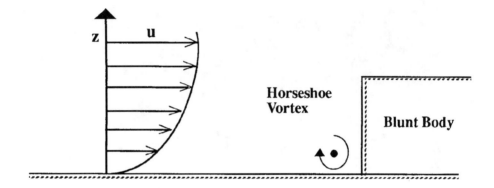

Figure 46. Shear flow in a corner.

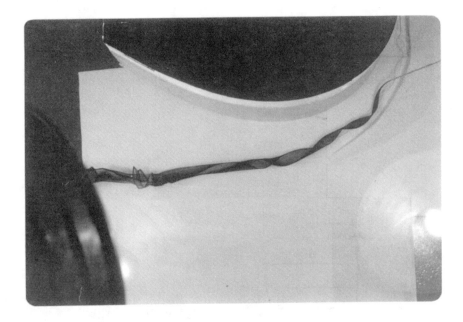

(a)

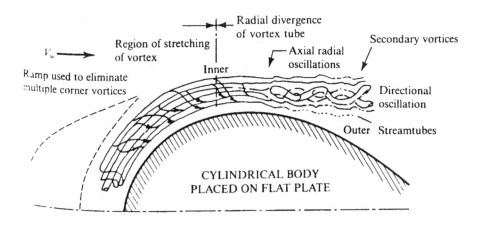

(b)

Figure 47. (a) Laboratory simulation of shear flow past a blunt body creating the horseshoe vortex. (b) Schematic of (a) showing various regions of flow in the horseshoe vortex.

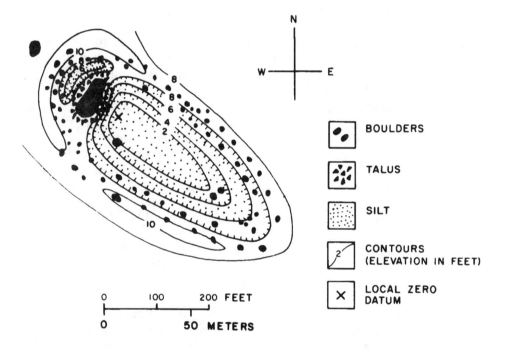

Figure 48. Scour hole development near the 18 m × 11 m × 8 m boulder (Figure 44). From THE CHANNELS OF MARS by Victor R. Baker, Copyright © 1982. By permission of the University of Texas Press.

The rotational beating of the secondary vortices is a function of the circumferential Reynolds number which is equal to the circulation of the primary vortex divided by the kinematic viscosity of the fluid. The greater the free stream velocity, the greater is the circulation and the smaller is the diameter of the primary horseshoe vortex.[20] However, the greater the free stream velocity, the greater is the "beater" speed of the secondary vortices. The behavior of vortices has been studied by Granger [62], and experimental set-ups are found in [63, 64].

A result of scouring can be seen in Figure 49. One clearly sees a scoured channel emerging from a rather complex terrain. The details of the scouring cannot be seen due to the great distances involved and due to erosion over millions of years, but surely the totality of the scouring and erosion has resulted in the relatively smooth channel.

Another form of erosion is due to vortices whose axis of rotation is nearly vertical. This type of vortex does not scour like the longitudinal vortex breakdown but "plucks" the material from the channel wall. One of the best explanations of this action is given by Bretz, Smith, and Neff [65]. Typical scabland erosion features like pot holes, column erosion, etc. are produced only by high discharge rates and steep gradient erosion. Figure 50 shows the erosion produced by secondary flow of vortices. Some call these vortices kolks; however, it is important to understand kolks are produced by tides, and on Mars, it is not appropriate to refer to these vertical vortices as being kolks. Vertical vortices causing serious bed and side wall erosion are due to fast-moving currents. A flow moves past an upper circulation, resulting in a sheet of fluid rolling up into a strong vertical vortex shown in Figure 51. Such vertical vortices hollow out the surface bed as the secondary flow forces material from the bed and walls upward into the main flood waters. This updraft can be explained by Bernoulli's equation of hydrodynamics.

Vertical vortices can also be formed by the sudden backup of an oscillatory turbulent boundary layer. A violent upward motion in the flooding might have occurred, resulting in concentrated localized spinning at the surface of the current. (See Jackson [66]).

Other types of localized vortex action stem from flow separation behind obstacles or from widening of the channel. These vortices are the familiar wake vortices. The strength of these vortices are functions of sedimentation content (or density differences). Caves and rock passages can create longitudinal vortices that can turn and become vertical vortices that can either remain fixed or move with the current.

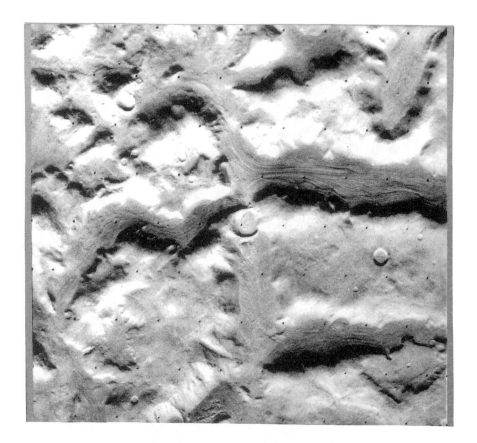

Figure 49. Mariner 9 photograph of the scouring of a channel on Mars. Courtesy of NASA.

Figure 50. Kolks formed by periodic vortices at the Irish coast. From VORTICES IN NATURE AND TECHNOLOGY, by Hans Lugt, Copyright © 1983. By permission of John Wiley and Sons, Inc.

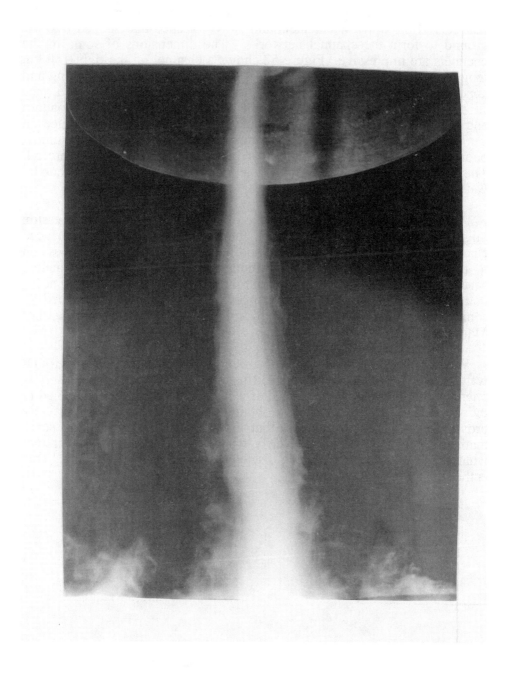

Figure 51. A columnar gas vertical vortex. Carbon dioxide is the gas. Air moving from left to right moves past a flush-mounted pool of dry ice in water. Over head are vanes at specific radial locations to generate upper wind circulation. The net result is a small-scale tornado.

We have discussed a number of mechanisms that could release the Martian flooding, and we have discussed the morphology and hydrodynamics of the flooding. We have addressed what happened to the water that carved the outflow channels and slowly seeped and sapped into the ground to form the channel networks. The dimensions of some of the channels are impressive. Figure 52 shows a channel on the right 150 km long and 5-6 km wide. Masursky [50] points out that the meandering and dendritic form of such a channel is strong evidence that some fluid once flowed through this channel. Note the similarity of the dendritic pattern and meandering with that of the scabland cataracts in Moses Coulee on earth shown in the upper right of Figure 53. An enlargement of a stubby, poorly developed dendritic pattern of Figure 52 is shown in Figure 54. This pattern is suggestive of ground sapping or soil flowage rather than by water collected from rains.

Almost every channel ends at a closed depression. The depression might be a crater, trough, crack, or enormous basin. The flat-surfaced chasm of Figure 54 may have been modified and widened by the recession of the walls as a result of undermining, or it may represent a channel carved out by a huge flood which burst forth from the chaotic terrain at the bottom left. The unusual pattern suggests it could have been formed by some type of sapping rather than by run-off rain water from an earlier Martian period.

We discussed the plausibility of ground ice, the plausibility of ground water sapping, ice covered lakes, diffusion of water vapor to the atmosphere, and other possible storage sites as they pertain to what happened to the water on Mars. Until we can obtain actual fluid data to conclusively prove what transpired, the best that can be done is to generate some mathematical models that can verify what we already know and predict what we do not know but hope to discover when man lands on Mars. This is the subject of the next chapter.

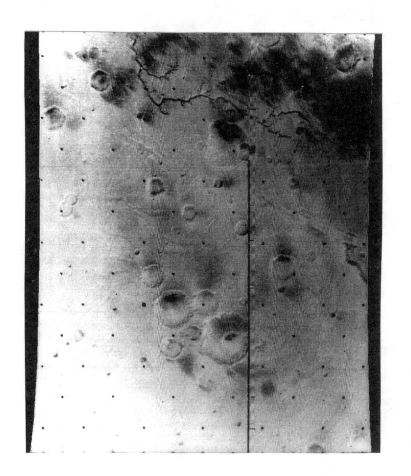

Figure 52. Mariner 9 photograph of a Grand Canyon type channel on Mars. The channel is 150 km long and 5 km wide. Photograph taken at an altitude of 1,655 km. Courtesy of NASA.

Figure 53. Satellite photograph of scabland cataracts in Moses Coulee. Courtesy of NASA.

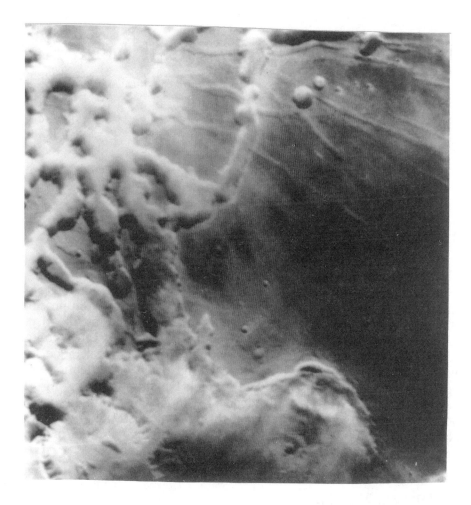

Figure 54. Mariner 9 photograph of a complex labyrinth composed of numerous intersecting closed depressions. Withint some of the depressions, faults in the surrounding plateau have near vertical dips. Courtesy of NASA.

8

MATHEMATICAL MODELS

8.1 BACKGROUND

It is very difficult, if not impossible to devise a mathematical model for the global distribution of water on Mars, since so many parameters play primary and secondary roles in the kinematics and equilibrium processes of the evolution of water. The unsteady boundary conditions, the physics of the ice, and lack of an appropriate equation of state for water are just a few of the problems. However, there are two mathematical models that stand out above all the rest and are deserving of special attention. One model will be identified as the FSZP model, Fanale et al. [67, 68]. The other model is the Wallace and Sagan [15] model, identified as the WS model. The former model treats the global distribution of water (subsurface ice) on Mars, and the latter treats the evaporation of ice on Mars. We shall treat the WS first and then the FSZP model.

8.2 THE WS MODEL

The energy flux to the upper ice surface is

$$(1 - A_i)q_i - \varepsilon_i \sigma T^4 + K_i \frac{\partial T}{\partial z} - L_S \dot{m}_u = 0 \qquad (42)$$

and

$$A_i q_i - K_i \frac{\partial T}{\partial z} + L_i \dot{m}_\ell = 0 \qquad (43)$$

for the <u>lower</u> surface, where A_i is the albedo, q_i is the flux of sunlight absorbed within the ice, ε_i is the emissivity, σ is the Stefan-Boltzmann constant, and K_i is the thermal conductivity. The third term of Equation (42) represents the heat conducted. L_S is the latent heat of sublimation, L_i is the latent heat of fusion, $\dot{m}_u$ is the evaporation rate from the upper surface, $\dot{m}_\ell$ is the rate of freezing at the lower surface, and z is the vertical coordinate. As the ice sheet approaches equilibrium,

$$\dot{m}_u = \dot{m}_\ell \qquad (44)$$

characterizing a constant ice thickness.

The thermal conductivity of ice K_i, based on data from Ratcliffe [70] is

$$K_i(T) = (1.863/T) - 0.00147 \text{ cal } \text{cm}^{-1}\text{s}^{-1}(\text{K})^{-1} \qquad (45)$$

The fourth term in Equation (42) is the energy of evaporation. The mass flux $\dot{m}_u$ (g cm^{-2} s^{-1}) depends on three factors: the atmospheric pressure p_a, the wind velocity, and free convection.

8.2.1 Evaporation Rate Due to Pressure

Consider the effect of atmospheric pressure on evaporation. If the vapor pressure p_v is greater than atmospheric pressure p_a, it is simple to show from the kinetic theory of gases that the mass flux is

$$\dot{m}_u = \alpha(p_v - p_a)(M/2\pi RT)^{1/2} \qquad (46)$$

where M is the molecular weight of the vapor and α is the coefficient of evaporation given by Tschudin [71] as

$$\alpha = 0.94 \pm 0.06 \qquad (47)$$

If $(p_v - p_a) < 0$, then vaporization is controlled by diffusion. This will be discussed next.

8.2.2 Evaporation Rate by Wind Velocity

The WS model considers evaporation by wind or, more properly, the atmospheric boundary layer of CO_2. Assuming a turbulent flow, $R_\ell > 5 \times 10^5$, we treat the ice surface as being essentially a flat plate. As shown in Figure 55 (Granger [60]), the turbulent boundary layer of the wind is partitioned into four regions: a viscous sublayer, a log law region, an outer-law region, and a viscous superlayer. The first two lower regions next to Mars' surface are called the inner layer, and the last two outer regions are called the outer layer. The viscous sublayer is subdivided into a linear sublayer and a buffer layer. Granger [60] shows the dimensionless velocity u^+ in the linear sublayer as

$$u^+ = \frac{\bar{u}}{u_*} = \frac{u_* z}{\nu} = z^+, \ z^+ \le 3 \tag{48}$$

where

$$u_* = \sqrt{\frac{\tau_0}{\rho}} \tag{49}$$

$\bar{u}$ is the average fluid particle velocity in the x-direction (flood's direction), u_* is the friction velocity, ν is the kinematic viscosity equal to the dynamic viscosity μ divided by the density ρ

$$\nu = \mu/\rho \tag{50}$$

and z is the vertical height above the ice. We see from Equation (48) that the thickness t of the linear sublayer is

$$t = \frac{3\nu}{u_*}. \tag{51}$$

In the buffer zone, $3 < u_* z/\nu \le 40$, both viscous and turbulent shear stresses are important. The dimensionless velocity u^+ is

$$u^+ = 5\ln z^+ - 3.05 \tag{52}$$

and the thickness t of the buffer layer is

$$t = 40\,\nu/u_* \tag{53}$$

In the logarithmic law region, $40 < \frac{u_* z}{\nu} \le 0.2\,\delta\frac{u_*}{\nu}$, the dimensionless velocity is

$$u^+ = \frac{1}{0.41}\ln z^+ + 5.0 \tag{54}$$

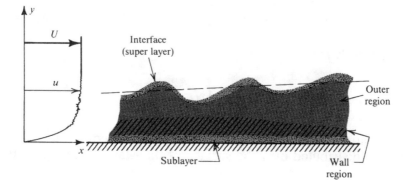

Figure 55. Regions of the wind turbulent boundary layer.

and the thickness t measured from the ice surface is

$$t = 0.2\delta \tag{55}$$

where δ is the turbulent boundary layer thickness, a function of Reynolds number. As an estimate, Granger [60] shows

$$\delta = 0.37\left(\frac{\nu}{Ux}\right)^{1/5} \tag{56}$$

where x is measured from the wind's stagnation point on the ice. In all of this, we have assumed the ice is relatively smooth, not an unreasonable assumption. The WC model states that if the friction velocity u_* is greater than the critical friction velocity u_{*_c}, where

$$u_{*_c} = 2.5\,\nu/z_o \tag{57}$$

then the wind profile is

$$u^+ = \frac{1}{0.41}\ln(z/z_o) \tag{58}$$

rather than Equation (54), where z_o represents the modeled average height of protrubances distributed over the surface of the Martian ice; $(z_o \sim 0.1$ cm).

The evaporation rate formula from the upper surface given by WS is of the form

$$\dot{m}_u = \Delta\rho\Big/\big\{(t/d) + (1/0.41\,u_*)\ln\big[(H+z_o)/(t+z_o)\big]\big\} \tag{59}$$

where H is the vertical height water vapor diffuses (typical values are 1 m), $\Delta\rho$ is the difference in the water vapor density between $z = 0$ and $z = H$, and d is the diffusion coefficient of water vapor in a pure carbon dioxide atmosphere. WS use data from Schwertz and Brow [72] to obtain an expression for the diffusion coefficient

$$d = 0.1654(T/273.15)^{3/2}(1013\ \mathrm{mbar}/p)\mathrm{cm}^2\ \mathrm{s}^{-1} \tag{60}$$

where p is in units of millibars. The dynamic viscosity μ, which is important in finding magnitudes of t and laminar sublayer velocity u^+ is given for CO_2 by WS as

$$\mu = \left(0.002162T^2 + 3.771T + 172.01\right) \times 10^7 \text{ dyne s cm}^{-1}. \qquad (61)$$

The value of $\Delta\rho$ is the saturation vapor density at the surface of the ice <u>if</u> the CO_2 is free of water vapor.

8.2.3 Evaporation Rate by Free Convection

In the absence of wind, evaporation can occur due to the ground being saturated with water vapor, the atmosphere at the ground being "less dense than the overlying dry CO_2 atmosphere and is therefore dynamically unstable." Ingersoll [73] shows the evaporation rate from the upper surface as

$$\dot{m}_u = 0.17\Delta\rho\, d\left[\left(\Delta\rho/\rho\right)g/v^2\right]^{1/3} \qquad (62)$$

where $\Delta\rho/\rho$ is the ratio of the difference in density produced by the water vapor in the atmosphere to the CO_2 density itself; i.e.

$$\frac{\Delta\rho}{\rho} = \frac{\left(M_{CO_2} - M_{H_2O}\right)p_v}{\left[M_{CO_2}p_a - \left(M_{CO_2} - M_{H_2O}\right)p_v\right]} \qquad (63)$$

where the molecular weight of CO_2 is 44 and M_{H_2O} is 18. WS point out that Equation (62) is applicable if the thermal stratification is neutral. If the stratification is not neutral, then Equations (62) and (63) must be altered since the gas density is dependent upon temperature. Gierasch and Goody [74] showed "that the Martian atmosphere alternates between highly stable and highly unstable stratification daily." WS point out that the evaporation rate $\dot{m}_u$ of an ice sheet is not seriously affected by free convection for three reasons: (1) "the greater thermal inertia of ice as compared to soil would reduce the diurnal temperature variation of the top layer of the ice sheet;" (2) "the incident solar radiation which penetrates the ice layer would melt the bottom few centimeters of the ice during the day," which, of course, freezes at night. Thus there would be only negligible effects on the surface ice temperature; and (3) "the ice layer would probably have a higher albedo than the soil, further reducing the solar heating which leads to thermal convection."

Since wind and free convection influence evaporation, WS suggested adding their effects, with the evaporation rate being

$$\dot{m}_u = \Delta\rho/(t/d)+(1/0.41\,u_*)\ln\left[(H+z_0)/(t+z_0)\right]$$
$$+\,0.17\Delta\rho\,d\left[(\Delta\rho/\rho)\,g/v^2\right]^{1/3} \tag{64}$$

Equation (64) is simply the sum of Equations (59) and (62).

8.3 THE FSZP MODEL

Fanale et al. [67, 68] developed an analytical model which is identified as the FSZP model, that gives the history and distribution of water on Mars. It starts with a uniform ice configuration at the beginning of Mars' geologic history in the same manner as the CH model. The FSZP model quantitatively treats (1) obliquity and eccentricity variations, (2) solar luminosity, (3) variations in the perihelion argument, (4) albedo changes at high latitudes due to CO_2 changes, (5) planetary heating and heat transfer, (6) vertical temperature variations as a function of time (millions of years) and latitude, (7) atmospheric pressure variations over a 10^4 time scale, (8) influences on the polar cap temperature, and (9) diffusion (both molecular and diffusion) of water through the regolith (the same topic treated by the CH model).

8.3.1 Obliquity and Eccentricity Variations

Figure 26 (Ward [69]) shows the obliquity and eccentricity values on Mars versus time (in millions of years). The average value of obliquity is approximately 24°, a maximum amplitude about 13.6°, a maximum value of 38°, and a minimum value of 10.8°. The short term period is seen to be about 1.2×10^5 years. These oscillations are superimposed on other longer term periods of about 1.2×10^6 years.

FSZP show the obliquity variation $\theta(t)$ as a function of real time, vertical distance z, short-term period P_S and long-term period P_L

$$\theta(t) = \theta_{av} + \theta_o \sin\frac{\pi t'}{P_L} - 4\theta_o \sin\frac{\pi t'}{P_L} z$$
$$\cdot\left[\frac{1}{4} - \frac{1}{\pi^2}\sum_{n=1}^{\infty}\frac{(1-\cos n\pi)}{n^2}\cdot\cos\frac{(\pi n t')}{P_S}\right] \tag{65}$$

where t' is the time from the beginning of a long-term period, θ_{av} is the average value of the obliquity, and θ_o is the maximum amplitude of the oscillation.

A similar expression to the obliquity is that for the eccentricity $e(t)$. Using Figure 26, the average eccentricity e_{av} is 0.073, the maximum eccentricity is 0.141. The short-term period P_S is 9.5×10^4 years, and the long-term period P_L is 2×10^6 years. FSZP describe the eccentricity as

$$
e(t) = e_{av} + e_o \sin\frac{\pi t'}{P_L} - 4e_o \sin\frac{\pi t'}{P_L} z
$$
$$
\cdot \left[\frac{1}{4} - \frac{1}{\pi^2} \sum_{n=1}^{\infty} \frac{(1-\cos n\pi)}{n^2} \cos\frac{2\pi n t'}{P_S}\right]
\tag{66}
$$

8.3.2 Temperature Boundary Value Problems

The temperature T of the regolith and surface is obtained from the well known thermal diffusion equation

$$
K\frac{\partial^2 T}{\partial z^2} + S_h - \frac{K}{\alpha}\frac{\partial T}{\partial t} = 0
\tag{67}
$$

where the first term is the heat transfer rate per unit volume by conduction, the second term is the heat generation rate per unit volume $S_h = 2.55 \times 10^{-7}$ ergs/s cm^3, and the last term is the unsteady heat transfer rate per unit volume. Here α is the thermal diffusivity equal to $K/\rho_S C_S$ (soil density and specific heat). We require two boundary conditions: one at the lower limit of the regolith ($z = L$) and one at the surface ($z = 0$):

$$
K_i\frac{\partial T}{\partial z}\bigg|_{z=0} = \varepsilon_i \sigma T^4 - (1 - A_i)q_i - L_S \dot{m} - F_a
\tag{68}
$$

$$
\frac{\partial T}{\partial z}\bigg|_{z=L} = -Q/K_i
\tag{69}
$$

where K_i is thermal conductivity of ice $= 8.0 \times 10^4$ ergs/cm·s·K, Q is the heat flux at $z = L$ ($Q = 29.33$ ergs/cm^2 s), A_i is the annual averaged radiometric albedo, q_i is the flux of sunlight absorbed within the ice, ε_i is the emissivity of ice (assumed unity), F_a is the downgoing infrared flux from the atmosphere (2% of maximum solar flux), L_S is the latent heat of sublimation, and $\dot{m}$ is the evaporation rate. The flux q_i is often evaluated using

$$q_i = \frac{S_S}{R^2}\cos i \qquad (70)$$

where S_S is the solar constant, R is the distance from Mars to the sun, and i is the angle of incidence at the surface. It is important to note Equation (69) is not applicable to either polar cap.

8.3.3 Pressure Field

The pressure field $p(z,t)$ is obtained by integrating the capillary diffusion equation (given by Fanale et al. [75])

$$(E+\beta)\frac{\partial p}{\partial t} = 4\frac{E^2 kr}{3}\left(\frac{2RT}{\pi M}\right)^{1/2}\frac{\partial^2 p}{\partial z^2} \qquad (71)$$

where the absorption parameter β is defined as

$$\beta = \partial p_a/\partial p\big|_T, \qquad (72)$$

p_a being the equivalent pressure of the absorbed molecules,

$$p_a = \rho_a RT/M, \qquad (73)$$

E is the porosity, k is a geometric factor to account for nonuniformity of actual pores, and r is the mean radius of an equivalent set of capillaries. Knowledge of the atmospheric pressure distribution is required to determine the time and latitude of the carbon dioxide polar cap and to calculate water fluxes through the atmosphere and regolith. FSZP calculated the CO_2 atmospheric pressures at each point in the obliquity cycle, assuming the regolith's material was basalt.

8.3.4 Velocity Field

The velocity field $\vec{V}$ is characterized by its three scalar components, u, v, w, in the x, y, z directions, respectively, as

$$\vec{V} = u\vec{i} + v\vec{j} + w\vec{k} \qquad (74)$$

The steady state Boussinesq equations for a compressible atmosphere with Coriolis terms added to the linear momentum equations are:

$$v\frac{\partial u}{\partial y} + w\frac{\partial u}{\partial z} - fv = 0 \qquad (75)$$

$$v\frac{\partial v}{\partial y} + w\frac{\partial v}{\partial z} + fu + \frac{\partial(p/\rho)}{\partial y} = 0 \qquad (76)$$

The continuity equation is

$$\frac{\partial(\rho u)}{\partial x} + \frac{\partial(\rho v)}{\partial y} + \frac{\partial(\rho w)}{\partial z} = 0 \qquad (77)$$

Equations (75)-(77) form a set of three homogeneous equations for u, v, and w. Note, $v = w = 0$ is a trivial solution, such that Equation (76) becomes

$$fu = \frac{\partial(p/\rho)}{\partial y} \qquad (78)$$

Note the above simple regime is in complete balance regardless how large the zonal velocity is, because the motion is exactly at right angles to the temperature gradient, i.e. $\partial(p/\rho)/\partial y$, and thus transports no energy. In actuality, there would be "some" non-zonal components of flow due to the viscous deceleration term $v\nabla^2 V$. This effect can be treated by examining the Taylor number

$$T_a = \left(\frac{fV}{v\nabla^2 V}\right) \qquad (79)$$

If T_a is large, then viscous deceleration is large, which is equivalent to the Coriolis effect fV being large. Typical values of Taylor number for Mars is 10^6, assuming $\nu = 10^5$ cm^2/s.

9

ORIGIN OF LIFE

9.1 BACKGROUND

In Chapter 2, we spoke briefly of the birth of earth 4.6 billion years ago: a lifeless place. A billion years later, the planet was filled with organisms, mostly algae. How did that happen? Could the same thing happen on Mars?

Each century seems to have its champions proposing explanations of how life began. Before 1650, most everyone believed God, or gods, created man (the highest form of life, as we know it, on earth). Each century after that had its critics. Critics who thought life was not accidental, or explained by mystical concepts, but who thought that a more scientific, provable, or logical explanation was the answer. That all living things had certain common denominators. The 20th century provided some of these answers.

Three rather obvious and provable similarities exist in all life forms. First, all living things have similar carbon-rich compounds. Another is that proteins found in organisms stem from a set of 20 standard amino acids. A third is that organisms carry their genetic information in the nucleic acids RNA and DNA and use the same genetic code: a code that specifies the amino acid sequences of all the proteins any organism requires. Hence, nucleic acids store genetic information that spell out the required proteins needed to sustain life and perpetuate its future offspring. How did these nucleic acids and proteins come into existence?

9.2 ACIDS AND PROTEINS

It is doubtful both the acids and proteins arose simultaneously at the same place and time. One needs the other and vice versa. Carl Woese and Leslie Orgel proposed that the nucleic acid RNA might have come first. They suggested that RNA could have developed the ability to link certain amino acids into proteins. DNA was assigned the role of carrier of the hereditary information, and it evolved from RNA. Since DNA was more stable than RNA, it was responsible for heredity factors. Orgel explains[21] that RNA came before proteins since Woese and he had difficulty explaining any situation in which proteins could self reproduce without nucleic acids.

A very simple experiment that one can perform in any university laboratory clarifies the chemical reactions that probably occurred on the early earth and, therefore, possibly on Mars, that produced amino acids. A flask filled with water (representing the oceans) is heated (representing the high temperature of the earth's crust). The water in the flask is converted to water vapor and is allowed to circulate in glass tubes, representing the air's current. The tubes connect to another flask that represents the earth's atmosphere (methane CH_4, ammonia NH_3, hydrogen H_2, and water vapor). The gases in the flask are then subjected to electrical discharges (representing lightning—which always accompanies volcanic eruptions) that cause the gases to interact. The mixed gases then pass through another tube to a condenser (which lowers the temperature) simulating the temperature at the upper atmosphere, forming liquid water (rain) that eventually fell back into the ocean. The experiment revealed that amino acids had formed in the lightning-gas reactions. One of these acids was glycine. For example, if one takes formaldehyde (2 parts hydrogen, 1 part carbon, 1 part oxygen) plus ammonia plus hydrogen cyanide, one obtains aminonitrile and water. The aminonitrile combines with water to give glycine plus ammonia. Note, the formaldehyde and hydrogen cyanide are simple compounds found in the earth's atmosphere. This may not have been the actual way it was, but it does point out it could have taken place, not only on earth, but possibly on Mars also. Orgel points out other scientific discoveries that support the notion that an RNA molecule preceded the protein synthesis.

But how did RNA come into existence?

9.3 ORIGIN OF RNA

A first requirement is that those organic compounds needed for life could not come from an atmosphere rich in oxygen (as it is today: 21%). On Mars, the atmosphere is largely nitrogen and carbon dioxide. Only 1½% is oxygen. As explained previously, we need an atmosphere rich in methane, ammonia, hydrogen, and water vapor. That soup fired up with lightning will produce RNA. What about a similar atmospheric soup on Mars? Since earth and Mars were formed at approximately 4.6 billion years ago, both planets being hot, both having oceans of water, both having electrical impulses, it is very likely amino acids were formed on both planets. Hence, it is strongly probable that RNA also existed on Mars. There are some scientists who dispute this scenario. They believe the atmosphere did not contain these gases in sufficient amounts. Another scenario for the origin of life is that life could have been delivered by meteorite dust, comets, and interstellar particulates. However, there is strong experimental evidence that amino acids needed for the manufacture of proteins, hence life itself, were abundant on earth. In fact, amino acids arise very easily in conditions where oxygen and oxidizing is minimal. Another scientist (Juan Oro at the University of Houston) discovered an abundance of adenine in a simple mixture of hydrogen cyanide and ammonia in a water solution. Adenine, like guanine, cystosine, and uracil, is one of four nitrogen-containing bases in RNA and DNA and one of the most essential biochemicals to help life get started.

The above compounds have been found to exist in space. Some have been found in meteorites. In fact, water, ammonia, hydrogen, cyanide, and formaldehyde are abundant in the interstellar dust masses where new stars have their genesis.

The next question is how RNA was created from such molecules. There are many recommendations, but one in particular serves to illustrate how it might have taken place. The form of a reactive nucleotide is, as we have presented it, composed of four nitrogen-based compounds: adenine guanine, uracil, and cytosine, that randomly come together as polymers, one of which acts as a catalyst. Free nucleotides then line up on the catalyst as well as on other polymers and join to form what is called a complementary strand. The catalyst and its complement then separate. The catalyst then begins making a complement of the complement, thereby producing a copy of itself. The catalytic copying of the catalyst and of its complement then assures a replication of both strands.

Some scientists feel that RNA cannot replicate itself without the aid of proteins, even though some have done so in their laboratory.

Some scientists believe the first replicating system was not organic. A. Graham Cairns-Smith of Glasgow proposed that positively charged ions in clay were the repository of genetic information. A great many hypotheses exist trying to explain the origin of life, and a full explanation may never come to pass; however, that possibility is believed to be remote. It remains to be seen if this explanation of the origin of life is exclusively earthen. There are no restrictions on the gravitational, electrical, or mechanical fields on what has been described. All the processes could have transpired on Mars. Rates may have been different, and ultimate forms different, but the chemistry should be similar on both planets as the chemicals involved are universal.

10

EVOLUTIONARY PROCESS

10.1 INTRODUCTION

Life, as we know it, is the result of fortuitous and contingent reactions of millions upon millions of events: some minor, some major. Writing space restricts any discussion of the millions of minor events. Let us consider just four major events that occurred on Earth. (1) About 530 million years ago, in the Cambrian Age, an enormous explosion took place that nearly wiped out all multicellular animal life on Earth. The genus Pikaia, a small swimming creature that possessed a rotochord (a dorsal stiffening rod) survived and was fossilized in a period after this explosion. It was the forerunner of humans. (2) The evolution of fin bones on lobe-finned fish gave a means of providing a strong axis that could support weight on land resulting in the fact that vertebrates could be terrestrial. (3) Sixty-five million years ago a meteor struck the Earth, exterminating the large reptilian animals that had dominated life for 100 million years. (4) About 2 to 4 million years ago, a small specie of primates evolved into an upright position in Africa. But the evolutionary processes are not as significant as the paleontological process. Though the earth is 4.6 billion years old, discovery of the oldest rocks shows solidification is 3.6 billion years old. In these rocks can be found bacteria and mats of sediment trapped by these cells.

Imagine, for 4 billion of the Earth's 4.6 billion years, life had been EXCLUSIVELY single celled animals. It has only been the last 600 million years that animals have been multicelled. All major stages in animal's multicelled structure transpired in a 70 million year period. Only in the past 500 million years do we see a progression of anatomical growth. That is, after the Cambrian explosion do we see the results of multicellular development. So many forms died out, the phyla being our most ancient ancestor. There are many questions that remain unanswered. Why select that specie of lobe-finned fish to develop into a vertebrate sea creature that sought land and air, became linked, then walked upright and became man? What was the process that selected man to dominate life? Why not ants, the eagle, the porpoise? They have societies, drives, and intelligence not too far removed from early man.

10.2 WHAT ABOUT MARS?

On Mars, we might expect to find fossilized life forms, perhaps no more than single celled animals. Discovering multicelled animals would be truly a fabulous discovery, as it would confirm the universality of life. However, chemical differences would alter the ultimate form of life on Mars. Finding any unicelled creature would be significant. For example, we might find liquid ammonia on Mars, resulting in a different form of biochemistry, forming different unicelled creatures. It may be liquid water exists at deep depths under Mars' surface and that bacteria survives in that dark, deep interior. In conclusion, the likelihood for life on Mars may appear dim, but all one needs is one specimen found in a fossilized state. And the evidence from the Mars meteor presented in Chapter 4 and shown in Figure 3-5 gives strong likelihood that life did and still may exist on Mars.

11

CONCLUSIONS

We recognize Mars is about 4.6 billion years old. At about 4.5 billion years ago, there was excessive melting and outgassing followed by solidification of a thin crust. The bombardment from extensive space debris declined, then the morphology stabilized into largely what we see today. The outflow and run-off channels date about 3.9 billion years ago and are presumed formed by catastrophic flooding of liquid water. The Martian atmosphere was surely warmer, more humid, and thicker than present. Considerable outgassing took place during the early melting. The thick atmosphere did not last too long, probably until 3.5 billion years old.

Much of the water from the run-off channels may have resulted in some vast aquifers that are below the permafrost which developed after the Martian atmosphere thinned. Periodic breakout of water formed enormous outflow channels, which have been found to be younger than the run-off channels. What happened to the water from the outflow channels is speculative. We suspect it did not return to the aquifer due to the ice layer sealing it off. Some of it may have drained back to the breakout, some may have vaporized, the rest may have been absorbed in the eolian debris or volcanic material, particularly in the relatively high latitudes.

We have indicated there is significant evidence of near-surface water, largely as ice, some possibly as liquid. There is ample evidence that a liquid eroded parts of the land producing streamlined shapes, troughs, islands, and scouring out channels. We should keep in mind that the surface of Mars is fixed, in contrast to the plate tectonic activity on earth Thus, the features on Mars have been stabilized for billions of years, exclusive of eolian erosion. When man lands on Mars, one hopes he will soon determine experimentally whether much of what others have postulated herein is true or not.

So what is ahead of us? On 4 July 1997, the U.S. Mars Pathfinder spacecraft is scheduled to land on Mars' surface. It will be surrounded by bags of air to soften the impact, but there is no question that the first impact will be severe, bounding into Mars' atmosphere about 150 feet. It is hoped to land in the Ares Vallis area, a flat, ancient flood plain. Once it comes to rest and it is still intact, a 22-pound robotic rover called Sojourner, about the size of a large bread basket, will be steered over an area the size of a football field. This area was chosen as it is believed the rocks on the plain's surface should be as old as ALH84001 and the search for fossilized life will begin.

After Pathfinder, the U.S. Mars Global Surveyor will orbit about Mars. It will contain measuring devices to do a detailed survey of the entire Martian surface and atmosphere.

After that, NASA wants to launch twin robotic spacecraft to Mars every 26 months until 2005. This plan of attack will provide a greater likelihood of partial success than if only one massive space launch were attempted. There are bound to be failures in the space program, but having a number of small probes should provide NASA with an enormous quantity of data even if there are failures.

U.S. Mars Orbiter '98 will have an advanced optical camera to search for sources of water in the Martian surface soil. U.S. Mars Polar Lander '98 is to examine the polar regions. Prior to entering Mars' atmosphere, two probes will be launched to study below-surface conditions.

Polar Lander will encounter ice. Ice is a great storehouse of the history of a planet, particularly early climate. Lander will dig troughs in the icy terrain in the south pole.

What can be done in 2000-2002 is too foggy to see any clear direction. Much depends upon the success of the earlier launches and their findings. Much depends upon the budget, much on the politics of the day.

We need to know that life exists on other masses than earth. How that search should continue is too debatable to resolve with any set of easy answers.

No one can deny that life on Mars has become a very hot topic. A poll[22] released recently showed that more than 50% of Americans believe that life exists outside of earth. With the hundreds of billions of stars and hundreds of billions of solar systems, that ours is the only one that sustains life is simply too simplistic for the rational mind. What that life form is or was or is to be depends upon an enormously complex set of factors involving energy levels, force fields, chemistry, kinematics, and thermodynamics. The life form could be as elementary as bacteria or as complex as an advanced intelligent being of the Star Trek mold.

So how should we progress to learn if life exists outside our planet? We can start with our closest planet: Mars. We can point to the Mars meteorite ALH84001 as good evidence[23] that life once existed on this cold, bleak planet. Is it worth spending hundreds of billions of American taxpayers' money to set humans on this planet, or are there simpler and cheaper ways to cement the issue of life or no life on Mars? We asked a similar question in Europe back in the 15th century. It took the financial resources of a kingdom to bankroll the explorers, and look what we gained from those ventures.

NOTES

(2) These weren't discovered until Mariner 9.

(3) The best presentation on what is needed for life to exist is found in Sir Denys Wilkinson's Pegram lectures given at Brookhaven National Laboratory, February, 1989.

(4) The following material is based on selections from Incompressible Fluid Dynamics, by Robert Granger, USNA Press, 1975.

(5) The Shadows of Creation: Dark Matter and the Structure of the Universe, by Michael Riordan and David Schramm, W.H. Freeman and Co., 1991.

(6) "The Origin of the Atmosphere," by Helmut Landsberg, *Scientific American*.

(7) The mathematical model was developed by James Walker, Paul Hays, and James Kasting, of the University of Michigan.

(8) Bruce Chalmers proposed his ideas that stemmed as a result of research on the solidification of molten metals.

(9) Some geologists prefer the term valley system.

(10) This depends upon how we define run-off channels. There are some quasi-dendritic channels on the flanks of some volcanoes, for instance.

(11) Maybe. Squyres has a suspicion an early higher geothermal gradient may have played a major role. (Personal note.)

(12) It should be noted Mars has no known active layer.

(13) There is an equally large consensus that confined aquifers could have formed during the development of permafrost.

(14) The log scale on the diagram obscures this point.

(15) Thermokarst involves melting and collapse. Permafrost requires only that $T < 273.15$ K: no ice content is involved. Terrain softening does not involve melting.

(16) In both cases, the cause is deformation of the ice that cements the debris.

(17) Some geophysicists refute this.

[18] The thickness will tend to be the same magnitude as the roughness elements; the real assumption is that the roughness elements are approximately 1 m, the value at least in the summer in Antarctica.

[19] The loss rate was probably higher early in Mars' history if the atmosphere was warmer and contained more water than it presently does.

[20] To simulate horseshoe vortices in the laboratory and study their behavior, one should work with low flow velocities in order to magnify the primary vortex diameter and hence enlarge the vortex breakdown.

[21] Leslie Orgel published a magnificent article on the origin of life in *Scientific American*, pp. 77-83, October, 1994. A detailed scientific explanation of what is briefly contained in the present book is presented in this very clear and exciting paper.

[22] Louis Harris and Associates poll released 2 September 1996.

[23] Some scientists, such as Professor Edward Scott at Hawaii's Institute of Geophysics and Planetology, believe the alleged traces of life were found not by ancient organisms but by the huge shock wave that sent the rock hurtling into space millions of years ago. He proposed the carbonate grains were formed from a hot, highly pressurized fluid that jetted into fractures in the surrounding rock. Some researchers believe that the grains were created during impact on earth. Unfortunately, Scott and his associates did not possess in their sample of the rock the carbonate globs shown in Figure 4 containing the remnants of these ancient microbes. (AP release 22 May, 1997)

REFERENCES

1. Squyres, Steven W., "Urey Prize Lecture: Water on Mars," ICARUS 79, 229-288 (1989).

2. Squyres, Steven W., "The History of Water on Mars," *Ann. Rev. Earth Planet. Sci.,* 12, 83-106 (1984).

3. Carr, M.H., "Formation of Martian Flood Features by Release of Water from Confined Aquifers," *J. Geophys. Res.,* 84, 2995-3007 (1979).

4. Baker, V.R., "Erosional Processes in Channelized Water Flows on Mars," *J. Geophys. Res.,* 84, 7985-93 (1979).

5. McCauley, J.F., Carr, M.H., Cutts, J.A., Hartman, W.K., Marsursky, H. et al., "Preliminary Mariner 9 Report on the Geology of Mars," ICARUS, 17, 289-327 (1972).

6. Marsursky, H., Boyce, J.M., Dial, A.L., Schaber, G.G., Strobell, M.E., "Classification and Time of Formation of Martian Channels Based on Viking Data," *J. Geophys. Res.,* 82, 4016-38 (1977).

7. Milton, D.J., "Carbon Dioxide Hydrate and Floods on Mars," *Science,* 183, 654-56 (1974).

8. Sharp, R.P., Malin, M.C., "Channels on Mars," *Geol. Soc. Am. Bull.,* 86, 563-609 (1975).

9. Nummedal, D., "The Role of Liquification in Channel Development on Mars," NASA TM 97929, 257-58 (1978).

10. Nummedal, D., "Debris Flow and Debris Avalanches in the Large Martian Channels," NASA TM 81776, 289-91 (1980).

11. Lucchitta, B.K., Anderson, D.M., Shoji, H., "Did Ice Streams Carve Martian Outflow Channels?," *Nature,* 290, 759-63 (1981).

12. Malin, M.C., "Age of Martian Channels," *J. Geophys. Res.,* 81, 4825-45 (1976).

13. Soderblom, L.A., West, R.A., Herman, B.M., Condit, C.D., "Martian Planetwide Crater Distribution, Implications for Geologic, History and Surface Processes," ICARUS 22, 239-63 (1974).

14. Pieri, D., "Martian Valleys: Morphology, Distribution, Age, and Origin," *Science,* 210, 895-97 (1980).

15. Wallace, D. and Sagan, C., "Evaporation of Ice in Planetary Atmospheres: Ice-Covered Rivers on Mars," ICARUS, 39, 385-100 (1979).

16. Sharp, R.P., "Mars: Fretted and Chaotic Terrain," *J. Geophys. Res.,* 78, 4073-83 (1973).

17. Czudek, T. and Demek, J., "Thermokarst in Siberia and its Influence on the Development of Lowland Relief," *Quat. Res.*, 1, 103-120 (1970).

18. Hussey, K.M. and Michelson, R.W., "Tundra Relief Features near Point Barrow, Alaska," *Arctic*, 19, 162-184 (1966).

19. Farmer, C.B. and Doms, P.E., "Global Seasonal Variation of Water Vapor on Mars and the Implications for Permafrost," *J. Geophys. Res.,* 84, 2881-2888 (1979).

20. Kiefer, H.S., Chase, S.C., Miner, E., Munch, G., and Neugebauer, G., "Preliminary Report on Infrared Radiometric Measurements from the Mariner 9 Spacecraft," *J. Geophys. Res.,* 78, 4291-4312 (1973).

21. Fanale, F.P., Salvail, J.R., Zent, A.P., and Postawko, S.E., "Global Distribution and Migration of Subsurface Ice on Mars," ICARUS, 67, 1-18 (1986).

22. Clifford, S.M. and Hillel, D., "The Stability of Ground Ice in the Equatorial Region of Mars," *J. Geophys. Res.,* 88, 2456-74 (1983).

23. Durham, W.B., Heard, H.C., and Kirby, S.H., "Experimental Deformation of Polycrystalline H_2O at High Pressure and Low Temperature Preliminary Results," *J. Geophys. Res.,* 88, 377-392 (1983).

24. Kirby, S.H., Durham, W.B., Beeman, M.L., Heard, H.C., and Daley, M.A., "Inelastic Processes of Ice I_h at Low Temperatures and High Pressures, *J. Phys.,* 48, C1-227-C1-232 (1983).

25. Kirby, S.H., Durham, W.B., and Heard, H.C., "Rheologies of H_2O Ices I_h, II, and III at High Pressures. A Progress Report," in <u>Ices in the Solar System</u> (J. Klinger and D. Benest, Eds.), Reidel, Dordrecht (1985).

26. White, S.E., "Rock Glaciers and Block Fields; Review and New Data," *Quat. Res.*, 6, 77-97 (1976).

27. Thompson, E.G. and Haque, M.I., "A High Order Finite Element for Completely Incompressible Creeping Flow," *Int. J. Num. Meth.*, 6, 315-321 (1973).

28. Thompson, E.G., Lawrence, R.M., and Lin, F-S., "Finite Element Method for Incompressible Slow Viscous Flow with a Free Surface," *Dev. Mech.*, 5, 93-111 (1969).

29. Bridgeman, W.P., "The Phase Diagram of Water to 45,000 kg/cm," *J. Chem. Phys.*, 5, 964-6 (1937).

30. Hobbs, Peter V., <u>Ice Physics</u>, Clarendon Press, Oxford (1974).

31. <u>VIIth Symposium on the Physics of Chemistry of Ice</u>, 1-5 Sept 1986, Grenoble (France), *Journal de Physique* Tome 48, Colloque C1, supplement on n°3, Mars (1987).

32. Kirby, S.H., et al., <u>Ices in the Solar System</u> (Ed. Klinger, J. et al.), 89 Reidel, Dordrecht, Holland (1985).

33. Kirby, S.H. et al., *J. de Pys.*, C1-229 (1987) (see [36]).

34. Thomas, D.G., "Transport Characteristics of Suspensions. VIII. A Note on the Viscosity of Newtonian Suspensions of Uniformed Spherical Particles," *J. Colloid. Sci.*, 20, 267-277 (1965).

35. Haynes, D.F., "Strength and Deformation of Frozen Silt," Third Inter. Conf. on Permafrost, pp. 656-661 (1978).

36. Mouginis-Mark, P.J. "Water or Ice in the Martian Regolith? Clues from Rampart Craters Seen at Very High Resolution," ICARUS, 71, 268-286 (1987).

37. Kuzmin, R.O., Boabina, N.N., Zabakeva, E.V., and Shashkina, V.P., "The Structure of the Martian Cryolithosphere Upper Levels," <u>In Workshop on Mars Simple Return Science</u>, LPI Tech. Rpt. 88-07, p 108 (1988).

38. Kochel, R.C. and Piper, J.F., "Morphology of Large Valley on Hawaii: Evidence for Groundwater Sapping and Comparisons with Martian Valleys," Proc. 17th Lunar & Planet. Sci. Conf., Pt. I., *J. Geophys. Res.*, 91, n. B13, E-175-E192 (1986)

39. Lucchitta, B.K., "Water and Ice on Mars: Evidence from Valles Mariners," NASA Tech. Mem. 89810, 313-315 (1987).

40. Nedell, S.S., Anderson, D.W., Squyres, S.W., and Love, F.G., "Sedimentation in Ice-Covered Lake Hoare, Antarctica," *Sedimentology*, 34, 1093-1106 (1987).

41. McKay, C.P., Clow, G.D., Wharton, R.A., Jr., and Squyres, S.W., "Thickness of Ice on Perennially Frozen Lakes," *Nature* (London), 313, 561-562 (1985).

42. Brutsaert, W., Evaporation into the Atmosphere, Reidel, Dordrecht (1982).

43. Sharp, R.P., "Mars: Fretted and Chatoic Terrains," *J. Geophys. Res.,* 78, 4073-4083 (1973).

44. Weiss, D., Fagan, J.J., Steiner, J., and Franke, O.L., "Preliminary Observations of the Detailed Stratigraphy Across the Highland-Lowland Boundary; Reports of the Planetary Geology Programs," NASA Tech. Mem. 84211, 422-425 (1981).

45. Lucchita, B.K., "Geologic Map of the Ismenius Lacus Quadrangle of Mars," USGS Map 1-1065, Atlas of Mars, 1:5,000,000 Geologic Series (1978).

46. Nedell, Susan S., "Formation of the Layered Deposits in the Valles Marineris Mars," Reports of Planetary Geology and Geophysics Program 1986, NASA Tech. Mem. 89810, 316-318 (1987).

47. Farmer, C.B., Davies, D.W., Holland, A.L., LaPorte, D.D., Doms, P.E., "Mars: Water Vapor Observation from the Viking Orbiter," *J. Geophys. Res.,* 82, 4225-48 (1977).

48. Jakosky, B.M., Farmer, C.B., "The Seasonal and Global Behavior of Water Vapor in the Mars Atmosphere: Complete Global Results of the Viking Atmospheric Water Detector Experiment," *J. Geophys. Res.,* 87, 2999-3019 (1982).

49. Squyres, S.W., "The History of Water on Mars," *Ann. Rev. Earth Planet Sci.*, 12, 83-106 (1984).

50. Mars as Viewed by Mariner 9, NASA SP-329 (1974).

51. Hartman, W.K. and Raper, O., The New Mars, the Discoveries of Mariner 9, p 124-125, NASA SP-337 (1974).

52. Clark, B.C., Baird, A.K., Rose, H.J., Toulmin, P., Christian, R.P. et al., "The Viking X-ray Flourescence Experiment: Analytical methods and Early Results," *J. Geophys. Res.,* 82, 4577-94 (1977).

53. Dzurisin, D. and Blasius, K.R., "Topography of the Polar Layered Deposits of Mars," *J. Geophys. Res.*, 80, 3286-3306 (1975).

54. Kieffer, H.H., "Mars South Polar Spring and Summer Temperatures: A Residual CO_2 Frost," *J. Geophys. Res.*, 84, 8263-88 (1979).

55. Cutts, J.A., "Nature and Origin of Layered Deposits of the Martian Polar Regions," *J. Geophys. Res.*, 78, 4231-49 (1973).

56. Neugebauer, G., Munch, G., Kieffer, H.H., Chase, S.C., Miner, E.D., "Mariner 1969 Infrared Radiometer Results: Temperature and Thermal Properties of the Martian Surface," *Astron. J.*, 76, 719-27 (1971).

57. Milton, D.J., "Water and Processes of Degradation in the Martian Landscape," *J. Geophys. Res.*, 78, 4037-47 (1973).

58. Baker, V.R. and Milton, D.J., "Erosion by Catastrophic Floods on Mars and Earth," ICARUS, 23, 27-41 (1974).

59. Baker, V.R. The Channels of Mars, University of Texas Press, Austin (1982).

60. Granger, R.A., Fluid Mechanics, 665-698, Dover Publ., Inc., New York (1995).

61. Costa, J.E., "Paleohydraulic Reconstruction of Flash Flood Peaks from Boulder Deposits in the Colorado Front Range," *Geol. Soc. of Am. Bull.*, 94, 986-1004 (1983).

62. Granger, R.A., "A Steady Axisymmetric Vortex Flow, *J. Geophys. Fluid Dynamics*, 3, 45-88 (1972).

63. Granger, R.A. "A Steady Axisymmetric Vortex Flow," Expt. 33, Experiments in Fluid Mechanics, ed. R. Granger, p 212-219, Holt, Rinehart, and Winston, New York (1988).

64. Escudier, M., "Swirling Flow in a Closed Cylindrical Container," Expt. 34, Experiments in Fluid Mechanics, ed. R. Granger, 220-227, Holt, Rinehart, and Winston, New York (1988).

65. Bretz, J.H., Smith, H.T.U., and Neff, G.E., "Channeled Scabland of Washington: New Data and Interpretation," *Geo. Soc. of Am. Bull.*, 67, 957-1049 (1956).

66. Jackson, R.G., "Sedimentological and Fluid Dynamical Implication of the Turbulent Bursting Phenomenon in Geophysical Flows," *J. Fluid Mech.*, 77, 531 (1976).

67. Fanale, F.P., Salvail, J.R., Zent, A.P., and Postawko, S.E., "Global Distribution and Migration of Subsurface Ice on Mars," ICARUS, 67, 1-18 (1986).

68. Fanale, F.P., Salvail, J.R., Zent, A.P., and Postawko, S.E., "Distribution and State of H_2O in the High Latitude Shallow Subsurface of Mars," ICARUS, 67, 19-36 (1986).

69. Ward, W.R., "Present Obliquity Oscillations of Mars," *J. Geophys. Res.*, 84, 237-241 (1979).

70. Ratcliffe, E.H., "The Thermal Conductivity of Ice. New Data on the Temperature Coefficient," *Phil. Mag.*, 1, 1197-1203 (1962).

71. Tschudin, K., "Rate of Vaporization of Ice," *Helv. Phys. Acta*, 19, 91-102 (1946).

72. Schwertz, F.A. and Brav, J.E., "Diffusivity of H_2O Vapor in Gases," *J. Chem. Phys.*, 19, 640-646 (1951).

73. Ingersoll, A.P., "Mars: Occurrence of Liquid Water," *Science*, 168, 972-973 (1970).

74. Gierasch, P. and Goody, R., "A Study of the Thermal and Dynamical Structure of the Martian Lower Atmosphere," *Plant. Space Sci.*, 16, 615-646 (1968).

75. Fanale, F.P., Banerdt, W.B., Saunders, R.S., Johansen, L.A., and Salvail, J.R., "Seasonal Carbon Dioxide Exchange Between the Regolith and Atmosphere of Mars: Experimental and Theoretical Studies," *J. Geophys. Res.*, 87, 10215-10225 (1982).

INDEX